Patrick Moore's Practical Astronomy Series

For futher volumes:
http://www.springer.com/series/3192

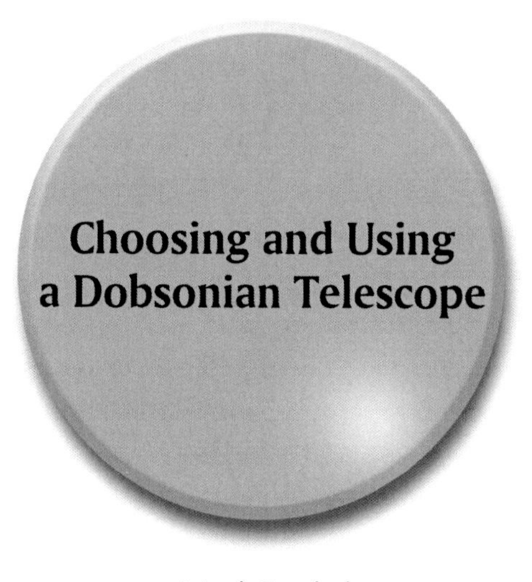
Choosing and Using
a Dobsonian Telescope

Neil English

Springer

Dr. Neil English
G63 0YB Glasgow
UK

ISSN 1431-9756
ISBN 978-1-4419-8785-3 e-ISBN 978-1-4419-8786-0
DOI 10.1007/978-1-4419-8786-0
Springer New York Dordrecht Heidelberg London

Library of Congress Control Number: 2011931711

Printed on acid-free paper

Springer is part of Springer Science+Business Media (www.springer.com)

The 1960s gave the world flower power, the contraceptive pill, the Moon rocket, and Billy Graham. But on the sidewalks of San Francisco, amateur astronomer John Dobson began an extraordinary evangelistic movement of his own, showing throngs of ordinary people how to build and use large aperture 'scopes from just about anything. Up until that time, large reflecting telescopes meant using heavy mirrors and complicated mounts, but Dobson almost single-handedly changed all that. By placing a large telescope on a simple, Lazy Susan mount, Dobson opened up large aperture observational astronomy to the masses. The Dobsonian revolution, as it came to be known, continues apace in the twenty-first century, with new and improved models appearing on the market every year.

Dobsonian-mounted reflectors marry simplicity of design with portability and large aperture prowess in ways that other telescope designs simply can't do. This book serves as the ultimate guide to buying and using a commercial Dobsonian for recreational astronomy. It provides in-depth accounts of the various models (plus accessories) on the market – both economy and premium – together with describing the wealth of innovations that amateurs have made to their Dobs to optimize their performance in the field.

There has been a huge increase in the popularity of these telescopes in the last few years, and Dobsonians (Dobs) have been heavily advertised in all the major astronomy magazines. In fact, they are now the best-selling large telescopes, both in Europe and the United States. Furthermore, if you happen to visit one of the many star parties taking place at different times of year around the world, you'll see a great variety of different Dob styles, ranging from the extravagant to the simplistic.

This book will be of particular interest to the many amateur astronomers who already have, or are intending to purchase, a Dobsonian telescope, perhaps to complement their existing arsenal of smaller 'scopes or to find out more about a premium model. But it is equally an attempt to attract enthusiasts of other telescope genres – refractors and catadioptrics, for example – to the great benefits these 'light buckets' can confer, as well as the very high quality images even economy-priced models can produce. Sadly, many thousands of Dobs lay dormant in the homes of amateurs around the world, neglected perhaps, for their portliness in comparison to a new ultraportable refractor. In this capacity, the goal of this book is to re-engage with this subset of 'lost amateurs' by expounding the great virtues Dobsonians already possess.

The book is arranged into two sections, for convenience. Part I recounts the story of John Dobson and the extraordinary movement he championed, before delving into the contemporary market. We will be showcasing everything from the very large to the very small and have endeavored to cover as many telescope models as possible. Due to space constraints, however, some products receive little or no mention. Our apologies in advance if we've not covered yours!

In Part II, we'll be dipping into the wealth of accessories that bring out the best performance of your Dobsonian, together with looking at ways to evaluate their optical performance under the stars. In this section too, we'll also be exploring how the humble Dob has now entered the twenty-first century, as auto-tracking and even Go-To models are now readily available, allowing amateurs to do things that were inconceivable only a few years ago. As you'll discover, things move pretty fast in the Dobsonian world!

May 2011

Dr. Neil English
Fintry, Scotland, UK

Acknowledgements

Any work of this nature could not have been accomplished without the help of a great many people. My gratitude is extended to Norman Butler, Peter Read, Bob Royce, Tom McCarthy, Scott Holland, Dave Bonandrini, Dave Rogers, Pat Conlon, David Hartley, Part Veispak, Javier Medrano, Vic Graham, Steve Dodson, John Compton, Normand Fullum, Doug Reilly, Kerry Robbert, Chris Hendren, Andy Sheen, Shane Farrell, Andre Hassid, Robert Katz, Jonathan Usher, and Dan Price.

I would also like to thank John Watson and Maury Solomon for endorsing the project and all the editorial team at Springer for a job well done. I would also like to thank my family for putting up with my seemingly endless retreats to the office to get this work done on time. Daddy's back!

About the Author

Experienced observer and writer on astronomy, Neil English has enjoyed looking through and writing about Dobsonian telescopes of all vintages. A Fellow of the Royal Astronomical Society, Neil has been a regular contributor to Britain's *Astronomy Now* for over 15 years. His work has also appeared in *Astronomy* magazine and *New Scientist.* He holds a PhD. in biochemistry and a BSc in physics and astronomy. English is the also the author of the sister Springer title, *Choosing and Using a Refracting Telescope.* He observes from the dark skies of rural, central Scotland, where he lives with his wife and two young sons.

Contents

PART ONE

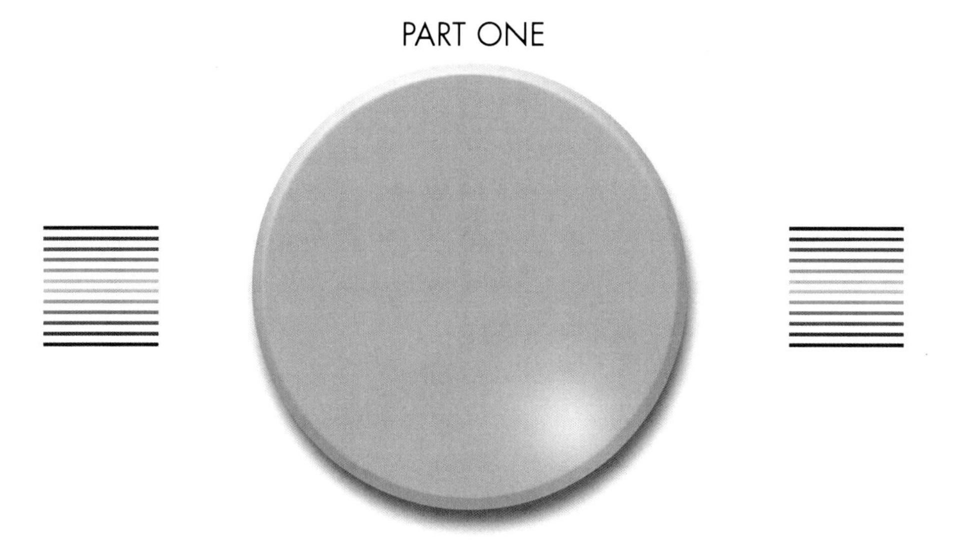

The Dobsonian Telescope

CHAPTER ONE

John Dobson, The Man and His Legacy

All great movements, whether political or cultural, begin with the individual. This is particularly apt when considering the legacy of John Dobson, who, almost singlehandedly started a revolution in amateur astronomy that has steadily gained momentum over the past 30 years.

John Lowry Dobson was born to American parents in Beijing, China, on September 14, 1915. His maternal grandfather was the founder of Peking University. His father was a lecturer in zoology at the university and his mother a musician. By the time John entered high school, China was experiencing considerable political unrest, and in 1927, the entire Dobson family (three brothers included) packed up and returned to the United States. John's father got a job teaching high school science, retiring in the 1950s. After completing high school, John read chemistry at the University of California at Berkeley, receiving his bachelor's degree in chemistry and mathematics in 1943. Like many of his fellow graduates, Dobson had to forego any chance of continuing his studies while the war in Europe and the Pacific raged on. Like so many bright young graduates, he soon found himself working on the Manhattan Project.

But Dobson quickly learned that such tasks were unsuited to his intellectual inclinations. His life changed forever when, out of sheer curiosity,

N. English, *Choosing and Using a Dobsonian Telescope*, Patrick Moore's Practical Astronomy Series 1, DOI 10.1007/978-1-4419-8786-0_1,

he attended a meeting at the Vedanta Center in San Francisco and was so bowled over by what he heard that he immediately enlisted and became a trainee of the Ramakrishna order. That was the way it would stay for Dobson for the next 23 years. While in the monastery he was assigned to reconciling the conflict between science and religion. In order to do that, he had to sample the wider universe for himself, which first led him to think about telescopes.

So in 1956, the 41-year-old holy man scraped together the materials to make a tiny 2-inch refracting telescope with a focal length of 14 inches delivering a magnification of 37×, from material he had scrounged from junk stores. With this he spied the minute Saturnian globe and its spellbinding system of rings. Those first glimpses had a profound effect on him. Indeed, it was as if he had some kind of religious experience. Only when he exhausted the limited power of the 2-inch glass did Dobson begin to pine for more light-gathering ability. A fellow monk at the monastery suggested he make his own mirror to satisfy his aperture fever. And dutifully, Dobson did just that, grinding it out of a piece of 120-inch 'marine-salvage' porthole glass.

Although the telescope was held together on a shoestring budget, the first look at a gibbous Moon through it changed his life forever and sent him headlong towards the evangelistic amateur astronomer and telescope maker he was to become. He brought his telescopes out onto the streets of San Francisco, for example, and began to preach the gospel according to John Dobson. He'd set up his ragtag Newtonian on the sidewalks of the city and cordially invite passersby to come take a look through his telescope. Gentle persuasion was his way of doing things.

From the very beginning, it seems, Dobson wanted everyone to witness the cosmic apparition that he himself encountered on that fateful evening in 1956. Two years later, Dobson had been transferred to another monastery in Sacramento. By then, he was churning out better and better telescopes. The first one was a little 50-inch reflector with a mirror ground from the bottom of a dug-out gallon jug. But soon he was busy grinding much bigger mirrors from free donations 'smuggled' into the monastery by appreciative fans. And he quickly became very good at it, too!

Grinding mirrors is a very time consuming activity, however, and public outreach demands even more. Soon, his superiors at the monastery grew concerned that it was eating too much into his monastic duties. Still, they tolerated his long AWOL spells from the monastery for many years to come. It was only in the spring of 1967 that the great man was asked to leave the monastery on the grounds that he could not fulfill his duties and continue playing the sky-watching evangelist. So after 23 years of living the Spartan existence of a monk, Dobson left in search of a new life. Predictably enough, he decided to dedicate the rest of his career to public astronomy outreach.

The type of telescope that bears his name was not patented. "That would be like trying to patent a cup with a handle on it," Dobson once remarked. His goal was to create a telescope that had good optics simply mounted. Equatorial platforms were far too complex for the task, so he settled for a simple Lazy Susan design in which the 'scope sits in a cradle that allows it to move up and down as well as from side to side. Actually, Dobson did not invent this mounting scheme, either. If you look at some old photographs from Stellafane dating back to 1941 – before the great man ever looked through a telescope – you can clearly see a few Newtonian reflectors that are mounted in a configuration that is remarkably similar to the 'Lazy Suzan' mounts found on contemporary Dobs (Figs. 1.1 and 1.2).

Dobson did, however, wish to reduce the cost of assembling a decent telescope and so resorted to the simplest mount he could put together on a minimalist budget. A simple cradle alt-azimuth mount was by far the simplest option available. By 1968, some of the folk John had guided and inspired started a public-service organization named the San Francisco Sidewalk Astronomers. At first, only fairly small 'scopes were used, but as the organization grew, larger telescopes were made and hauled out onto the streets. By 1970, the Sidewalk Astronomers had a 24-inch telescope that could be transported to any location. Soon they were bringing their light buckets to dark sky sites across the United States to every major star party. And at nearly every one, a big Dob looked skywards. In 1978, he was invited to Hollywood, not to star in a movie but to lecture adoring crowds and teach telescope making. And he did that faithfully for a further 26 years. Nobody knows exactly how many telescopes the man built, but it's probably of the order of several thousand.

As a teacher of telescope making, he could be impatient, even rude, so eager was he to impart the proper mirror grinding skills to his students. His methods of testing the figure of mirrors using a light bulb were also crude, but he was never aiming for perfection, just adequacy.

Dobson's ideas about the wider world are eccentric, even today, to put it mildly. You only need to view a few YouTube clips of his lectures to see what we mean.

Dobson has also tried his hand at book writing. In 1991 he authored *How and Why to Make a User-Friendly Sidewalk Telescope* with his editor, Norman Sperling. This very influential book helped popularize

Fig. 1.1. Amateurs standing behind an alt-aziumth-mounted Newtonian at Stellafane, 1941 (Image credit: Stellafane).

the Dobsonian mount that we know today. It also provides a mine of information about Dobson's own background and his belief in the importance of popular access to astronomy for proper appreciation of the universe. It even delves into his belief in a steady state universe. John Dobson, just past his 95th trip round the Sun at the time of writing, is as colorful and charismatic as ever.

Fig. 1.2. Another picture of the same telescope in the foreground from Stellafane 1941 (Image credit: Stellafane).

The Revolution Unfolds

Revolutions begin slowly and gain momentum before impacting the world. So it was with Dobson's evangelism. Although his mission to get everyone making telescopes on a nickel and dime budget appealed to many, others were slow to warm to the movement. Back in 1969, the former editor-in-chief of *Sky & Telescope* magazine, Charles A. Federer, doubted that Dobson's simple plywood mountings and light-bulb testing procedures would be of any lasting value to serious observers. But Federer misunderstood Dobson's point and underestimated his influence. Dobson wanted to show that practically anybody could build, transport, and use a big 'scope.

The first company to adopt the Dobsonian as a viable commercial product was Coulter Optical, which entered the astronomy world in 1968. They were also the first company to standardize the f/4.5 focal ratio seen on many commercial Dobs. Back then, the company specialized in producing low cost, so-so quality parabolic mirrors. By getting into the Dob business they struck gold, quickly establishing a reputation for themselves for producing their inexpensive but optically adequate Dobsonian-style reflectors. By 1980, Coulter had become the first manufacturer to market the "Dobsonian" telescope much as we know it today. Their Odyssey series catered to those with a taste for small, medium, and large aperture.

All of these telescopes utilized a very simple 1 ¼ inch sliding focuser tube made using a sleeve with adjustable tension. The Odyssey 1, for example, was a basic 13.1 inch f/4.5 instrument that allowed many amateurs to enjoy impressive deep sky views for a remarkably low price. The first Coulter tubes had a large box shape at their lower extremity, with a trapdoor where the mirror sat in a sling. Unfortunately, that meant that the mirror had to be removed each time you moved the telescope. It wasn't exactly portable either, weighing in at 120 pounds. The design was later changed to the more common tube we see today, comprised of a particle board push-pull mirror cell. Its thickness was reduced from the standard 1:6 ratio to 1:13. This shaved almost 20 pounds off the total weight, resulting in a much more transport-friendly telescope (Fig. 1.3).

All the Odyssey 'scopes came with their characteristic painted Sonotubes and particle board boxes. And it's easy to see why the series became a hit almost overnight. Even the clumsiest amateurs were keen to try their hand at building their own 'scopes. Soon, to sate the aperture fever of the amateur community, the company starting offering a 10.1" f/4.5 Odyssey Compact telescope. Another model was their giant Odyssey II, which had a 17.5" f/4.5 optical tube. All of these were initially produced with the box at the lower end of the tube, but later Coulter redesigned them with a standard tube.

By 1984, the company began offering much more portable Dobs, such as the Odyssey 8. This made use of an 8-inch f/4.5 optical tube on a small but sturdy Lazy Susan mount. This was supplemented about a decade later with the Odyssey 8-inch f/7 mounted on the same cradle. This spurred the 1995 announcement of the Odyssey 8 Combo, a unique combination of two optical tube assemblies, one Dobsonian mount, and one eyepiece. Both tubes could be placed on the one mount to provide either wide field views or greater magnification and both had 1¼" helical focusers. Shortly before the company's demise a new Odyssey 2 was made available – a 16-inch f/4.5 telescope with a standard 1 ¼-inch helical focuser and eyepiece.

Coulter Optical enjoyed great success for the best part of quarter of a century, but finally folded late in 1995, not long after the death of the creator of the company. In 1997, Murnaghan Instruments once again made the Coulter line of telescopes available, but admittedly at somewhat

Fig. 1.3. An advertisement for the Coulter 13.1-inch Dob appearing in the June 1980 issue of *Sky & Telescope*.

lower quality. In 2001 Murnaghan brought the line to the end for the last time. By then, the big telescope manufacturers had gotten the message loud and clear, and all maneuvered to capture a piece of the market. Meade, Celestron, and Bushnell all produced a line of good, but relatively inexpensive, Dobs. Cheap labor costs for Chinese workers meant that mirrors of good optical quality could be produced to fit these telescopes. That said, the Dobsonian revolution was far from over, and new innovations in design were just around the corner, led, as ever, by amateur telescope makers (ATMs).

Variations on a Theme

Although the Dobsonian, in its original form at least, provides large aperture in a transportable format, these 'scopes were still very heavy and awkward to move. What's more, while closed tube designs were adequate for smaller apertures, other ATMers sought new ways of cutting weight. The first documented breakthrough came with the publication, in *Telescope Making* magazine, issue 17 in 1981, of Ivar Hamberg's truss tube alt-az telescope. This innovation was unlike anything that had come before. Although it still used Teflon on Formica bearings, Hamberg's truss tube was designed to be taken apart for transport. Of course, it was still considered to be a Dobsonian, but it really bore as much relationship to Dobson's design as a go cart does to a car.

As a result, almost all large modern alt-az mounted telescopes of over 12.5-inch aperture today copy this design. Ivar's article introduced the collapsible truss tube, allowing disassembly and transport to dark skies from urban areas. This opened up a whole new field of large aperture deep sky observers and a whole new trend. Following on from that, a great number of refinements were introduced by David Kreige, who put an enormous amount of original thought and effort into this design to make it relatively inexpensive and user friendly. Ivar's original blueprint has further evolved to some ultralight designs to reinforce this trend. Despite all of these radical new designs they are not called "Hambergians" or "Kreigian" but universally as "Dobsonians."

In 2003, Orion USA launched their innovative SkyQuest Intelliscope series of closed tube Dobs that came with a computerized object locator. Of course, neither the telescope nor the locator were anything new, but they did offer a new level of versatility to the Dob user. As we'll see in a later chapter, the object locator was coupled to high resolution encoders on both altitude and azimuth axes – with a database of 14,000 objects – that could be found by pushing the telescope to the desired object in the

Fig. 1.4. The Orion Intelliscope brought object locator technology to the Dob market (Image credit: Orion USA).

sky. This was followed in 2006, when Meade Instruments launched their groundbreaking new truss tube series, the Light Bridge Dobsonians, in a very economical package that undercut many of the traditional truss tube Dobmakers. Cleverly designed and incorporating inexpensive but good Chinese optics, Meade offered their line in 8- to 16-inch apertures (the 8-inch has since been discontinued). Orion USA followed up on that by launching their own rendition of the truss tube design, complete with object locator (Fig. 1.4).

The basic Dobsonian mount is, of course, of the non-motorized alt-az variety, and although some loved the simplicity of nudging the instrument along, others became frustrated at having to constantly move the telescope throughout an observing run. The answer, of course, was to build an equatorial tracking platform. Several companies took up the engineering challenge during the 1990s, and they have steadily improved over the years. For the first time, you could place a large aperture 'scope onto a mount that kept the object in view as it moved across the sky, thereby increasing the time spent actually observing.

Finally, perhaps the most exciting development has been the introduction of automatic alt-azimuth tracking and even GoTo technology to the commercial Dobsonian. Sky Watcher recently launched their innovative flex-tube Dobsonians with built-in, dual-axis motors. That was followed in 2010, when Orion USA announced that their Dobsonians could not only track objects but had full GoTo capability.

We'll be following these exciting stories in later chapters, but for now let's take a closer look 'under the hood,' as it were, of the most important feature of any Dobsonian – the Newtonian reflector that's at its heart. That's the subject of Chap. 2.

CHAPTER TWO

Know Thy Dob

Since the vast majority of Dobsonians on the market use Newtonian reflectors, it is fitting to begin this survey of this rapidly changing market with a discussion concerning the ins and outs of the reflecting telescope, so that you can get the most out of it and maintain it in tiptop condition. This chapter will discuss everything you need to know about how your Newtonian works. On our road to understanding, we'll be dipping in and out of history to record the key events that shaped the evolution of the Newtonian into its quintessentially modern form.

The basic design of the Newtonian reflector – so named because of its invention by Sir Isaac Newton – has hardly changed since it was first conceived by the great scientist in 1668. Instead of using a convex lens to focus light, Newton used a finely polished spherical mirror. Astronomers had known about the possibilities of parabolic mirrors since 1663, when James Gregory, an English mathematician, envisioned a reflecting telescope that would bounce light between two mirrors, one with a hole in it to allow light to reach the eyepiece. Of course, being one of Europe's finest mathematicians, Newton was well aware of the properties of parabolic mirrors that would in theory produce even better images, but methods to "carve out" a parabolic surface presented a practical problem beyond him at the time. That's why he settled on the less than perfect spherical

N. English, *Choosing and Using a Dobsonian Telescope*, Patrick Moore's Practical Astronomy Series 1, DOI 10.1007/978-1-4419-8786-0_2,

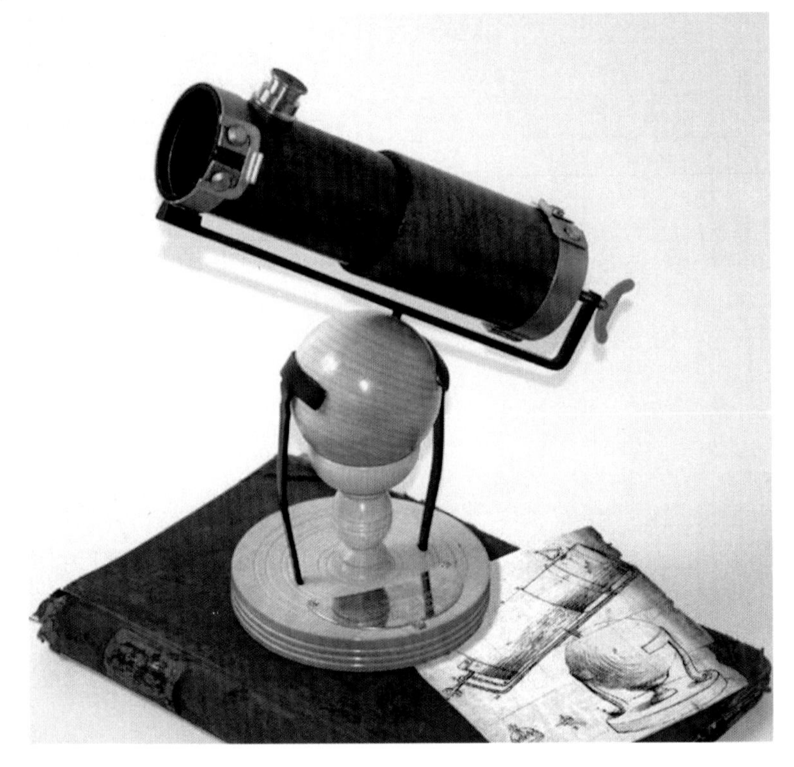

Fig. 2.1. A replica of Newton's reflecting telescope (Image credit: Pulsar Optical).

geometry for his metal mirror. The reflected light was sent back up the tube to a tiny flat mirror, mounted centrally and at a 45 degree angle, delivering the light cone to the eyepiece, where it reached focus. Newton was apparently very fond of pointing out that his little telescope – which delivered a power of about 40× – performed as well as refracting (lens-based) telescopes many times longer (Fig. 2.1).

Spherical mirrors are easier to make, but they have one minor flaw; light from the edges of a spherical mirror do not come to focus at the same point as rays from the center. In other words, the spherical mirror exhibits *spherical aberration,* which smears out the image so that it is difficult to get a razor-sharp view. That said, you can still obtain good results with spherical mirrors so long as the focal length of the 'scope satisfies the following formula:

$$\text{Focal length} = 4.46 \times (\text{Aperture})^{4/3}$$

This formula gives the minimum focal length a spherical mirror needs to be in order to meet the Rayleigh criterion, which is the lowest quality level that will produce an acceptably sharp image. For example, if you construct a 6-inch (15-cm) spherical mirror, it would need to have a minimum focal length of $4.46 \times (6)^{4/3}$. Plugging these numbers into a calculator gives a value of 48.6 inches (1,245 mm). There are commercially available Dobs that have spherical mirrors, but they are usually confined to apertures less than 6 inches for practical reasons.

When you take a mirror that has a nice spherical shape and deepen its curvature at the center a little bit, you will eventually arrive at a parabolic shape. It can be proven mathematically that only a parabolic surface has the attractive property of bringing to a single focus all rays parallel to its axis. In other words, a perfect parabolic mirror would have no spherical aberration. John Hadley, together with his two brothers, George and Henry, built the first reflector with a parabolic mirror, a 6-inch (15-cm) instrument of 62-inch focal length which he presented to the Royal Society in 1721.

For nearly two centuries after the invention of the Newtonian, the mirrors were made from a special alloy of mainly copper and tin. These "speculum" mirrors (practically 62 percent copper and 32 percent tin) gave a golden cast to the image and had a reflectivity of about 70 percent (actually a 1947 study suggested that its reflectivity varied from about 63 percent for blue light and 75 percent for red). After 6 months of exposure in a damp climate, its reflectivity drops by 10 percent, necessitating frequent polishing. Coupled to this, metal mirrors are exceedingly difficult to grind and are unduly heavy for their size. These deleterious aspects of speculum mirrors forced astronomers to look for better ways to build mirrors.

From Speculum to Glass

The next evolutionary step came in 1856, when the German astronomer Karl Steinheil, borrowing a technology developed by the chemist Justus von Liebig around 1840, hit upon a way of coating a 4-inch (10-cm) glass mirror with a thin veneer of silver. The telescope, by all accounts, gave excellent images but, remarkably, received little attention from the scientific cognoscenti. The following year, the physicist Jean Foucault made his own silver-on-glass mirror, and the resulting telescope – as well as the tests he singlehandedly developed to test its quality – received unanimous praise from the French Academy of Sciences.

These innovations set the scene for the rapid elevation of the reflecting telescopes in both the amateur and professional astronomy circuits that began in the middle of the nineteenth century and has continued unabated ever since. Glass mirrors could be made lighter and so more easily mounted inside their tubes. What's more, because parabolic mirrors work well, even at short focal ratios, they could be made much smaller than the standard instruments of the day – the long focus classical achromatic refractor – and thus were often more convenient to use in the field. But silver, despite its significantly greater reflectivity than speculum metal, was also subject to tarnishing over time. It was only after a series of influential experiments conducted in the 1920s that astronomers settled on the current reflecting material, aluminum. Although not nearly as reflective as silver, aluminum is far more durable and holds its shine for longer. All economically priced Dobs on the market have aluminized mirrors that reflect between 85 and 89 percent of the light striking them. With enhanced coatings however, reflectivities as high as 99 percent can be achieved (but, you guessed it, at extra cost).

Aluminum coating is done by depositing an ultra-thin layer of gaseous aluminum in a high vacuum tank. The aluminum metal is evaporated from a wire near the bottom of a purposefully built tank and coats the mirror which is rotated at the top of the tank. Then, a very thin silicon dioxide transparent coating is coated over the aluminum. During this process, the mirror never becomes hot.

Support Your Mirror

In order to perform well in the field, the mirror must be supported in its correct relative positions and orientations within the telescope tube. The earliest Dobs had their primary mirrors mounted inside simple sling-like devices that were OK but certainly not up to the task of holding collimation well after being lugged about in the back of a car. It goes without saying that the support must be firm enough to hold the mirror securely, but not so robustly as to stress the carefully figured shape of the mirror. The weight of the mirror is taken with supports on the bottom and sides. The mirror cell itself is housed inside a protective cell and held in situ by small clips (top supports), although smaller cells can be adequately supported using between 9 and 18 pads that provide structural support to the rear of the mirror. Lateral movement is kept in check by means of either a web-like sling or a teeter-totter post arrangement called a waffle tree (Fig. 2.2).

Fig. 2.2. A basic 9-point primary mirror cell (Image credit: John Heath).

The usual collimation arrangement for primary mirror cells is to have three adjustable spring tensioned bolts spaced 120 degrees apart. The primary mirror is aimed at the secondary mirror by adjusting these bolts up and down independently. There are a great number of primary mirror cells on the market, and they range from the barely functional to the highly elaborate. The most important function of this cell is to keep the position and orientation of the primary mirror securely fixed relative to the other optical components of the Newtonian.

The Secondary Mirror

The role of the secondary mirror, as previously mentioned, is to pick off the light gathered by the primary mirror and redirect into the drawtube, where an eyepiece can focus it. The secondary is usually mounted on spiderlike vanes attached to the side of the tube. The secondary is either a flat, elliptically shaped (because that's the geometry that minimizes the amount of light obstructed) mirror or a prism (Fig. 2.3).

Fig. 2.3. A typical secondary flat mirror attached to spiderlike vanes (Image by the author).

The latter are rarely used these days because they are only practical for smaller sized 'scopes less than 6 inches (15 cm). The elliptical mirror, which projects a circular geometry when viewed at the required angle, is carefully sized so as to minimize the amount of light cut off to the primary mirror but not so small so as to reduce the amount of light reaching the eyepiece (that would result in a darkening of the corners of the field of view and vignetting). Typically, central obstructions are expressed as a percentage of the diameter of the primary mirror. Short focal length Dobs usually have central obstructions of the order of 25 percent or greater, whereas longer focal length instruments (f/8 or above) can have central obstructions as low as 15 percent. Newtonian reflectors even of f/6 design can easily accommodate secondary mirrors less than 20 percent of the diameter of the primary mirror and still give full illumination to a reasonable field of view. (Indeed, this author has often entertained the idea of making a dedicated 6-inch f/6 planetary 'scope with 20 percent central obstruction. It might not fully illuminate wide field eyepieces, but that's not what this 'scope would be used for).

For focal ratios of f/8 and higher, secondary obstructions of 15 percent and smaller can be used productively. Compare that to compound reflectors of the Cassegrain and Gregorian types, either in classical or Schmidt Cassegrain or Maksutov configurations, which require secondary obstructions between 25 and 35 percent of the diameter of the primary mirror. In general, only the Maksutov Newtonian has smaller obstructions than the Newtonian.

A large central obstruction will make it more difficult to see subtle low-contrast detail, such as Jupiter's wispy equatorial festoons, or the delicate surface markings on Mars. The point at which this loss of image contrast becomes noticeable is a matter of heated debate, but most experienced observers agree that as long as the secondary mirror's diameter is less than 20 percent that of the primary mirror, its effects should be all but impossible to see. For binary star work some have claimed that a large secondary obstruction actually enhances performance, since it slightly shrinks the apparent size of the Airy disk, throwing the light into the first order diffraction ring and thereby somewhat increasing the apparent resolving power of the instrument. That one possible case aside, telescopes having central obstructions in excess of 25 percent begin to degrade the image noticeably in the arena of low-contrast resolution. When the size of the secondary reaches 30–35 percent, as it does in many commercial Schmidt Cassegrain telescopes, the drop in the ability to resolve fine detail becomes extremely noticeable. No matter how good your optical system may be, the mere presence of such a large secondary mirror may degrade the final wavefront to an effective quarter wave, PV, or less (Fig. 2.4).

Fig. 2.4. A typical secondary mirror housing for a small Dob (Image by the author).

Equally important is the quality of the secondary mirror itself. How flat is it? And how much light does it reflect? Standard "flats" come with a surface accuracy of about 1/10th wave, that is, they do not depart from perfect flatness by no more than about 50–60 billionths of a meter. With standard coatings they typically reflect about 96 percent of the light incident upon them. Of course, you can get flats that are considerably better than this, but at additional cost.

Some observers find the diffraction spikes (caused by the vanes of the secondary mirror) a distraction. Typically, four vanes are used, but some cut that down to three or have designed their own ingenious ways of mounting their secondary. Curved vane secondary spiders, which were first introduced back in 1931, virtually eliminate the diffraction spikes common to Dobs and Newts with straight vane spiders. But do curved spiders really lessen the diffraction at the eyepiece? No, they actually create more diffraction. Why? Well, for one thing, a straight vane causes diffraction over a 2 percent area. But a curved vane actually intercepts about 66 percent more area than straight vanes, so it will introduce more diffraction into the image.

Besides, if curved spiders were of any real advantage at all, professional observatories would be using them. Indeed, if your forte is teasing out the faint companion of a bright double star, the curved diffraction pattern will often bury it. In contrast, with a straight vane spider, you can always nudge the faint companion star out from under the diffraction spike and observe. Nor is there only one good way of doing this. Any departure from the straight four-vane design will give noticeably different results. Some companies, such as 'scopetronics and Protostar, even sell retrofit units that accommodate a wide variety of commercially made Dobs. Are they worth the extra expenditure (a $100–$200 venture)? If you've spent a lot of time around refractors, you may be tempted to do this. If you were weaned on reflectors, chances are you'll likely be sticking with convention. It's up to you and your tastes.

Image Quality

In the end, what really matters is the quality of the images your Dob serves up on celestial objects. The mirror, being freed from the nuisance of chromatic aberration (the main drawback in inexpensive achromatic refractors), serves up color pure images each and every time. But there's more to image quality than lack of false color. Here's a list of the aberrations that can degrade a telescope image.

Aberration	How They Scale
Spherical	$1/F^3$
Astigmatism	$1/F$
Coma	$1/F^2$
Distortion	$1/F$
Field curvature	$1/F$
Defocus	$1/F^2$

Let's tackle spherical aberration first. A perfect mirror focuses all incoming light to a sharp point on the optical axis, which is usually along the center of the telescope tube. However, a real mirror focuses rays more tightly if they enter it far from the optical axis than if they enter it close to the optical axis. This defect is called spherical aberration.

How does spherical aberration impair the image in the reflecting telescope? At low magnifications, little or no effects can be seen, but as you crank up the power an instrument displaying significant spherical aberration will be very hard to focus sharply. As a result, high power views of planets and the Moon take on a slightly "soft," drowned-out appearance.

Coma is an off-axis aberration. By that, we mean that stars in the center of the field are not affected, but the distortion grows stronger toward the edge of the field. Stars affected by pure coma are shaped like little comets (hence the name) pointed toward the center of the field. The effect is particularly common in reflecting telescopes, especially those that have fast focal ratios (f/3.5 to f/5). Fortunately, Al Nagler, founder of TeleVue optics, developed the Paracorr, auxiliary optics that do a great job correcting much of it. You slot the device into the focuser ahead of your eyepiece, much like a Barlow lens. The Paracorr extends the effective focal length of your Dob by 15 percent. Some observers have noted that while certainly correcting for coma it reduced on-axis definition just a little. Baader Planetarium (Germany) has developed another version of the Paracorr. Called the Multi-Purpose Coma Corrector (MPCC), it retains the native focal length of your 'scope. Sky Watcher has also introduced an economically priced coma corrector for their f/5 Dobs. We'll be delving into these magical devices in more detail in a later chapter.

Another aberration to look out for is astigmatism. This occurs when a mirror is not symmetrically ground around its center or, more usually, by misaligned optics. Most of the time, when such a system is misaligned or badly reassembled, slightly out-of-focus stars take on an oblate appearance. What's more, when you flip from one side of focus to another, the oval flips orientation by 90 degrees. In focus, images appear distorted, too.

The last two Seidel errors – distortion and field curvature – are in many ways less important. Field curvature is easy to spot. First, focus the star at the center of the field and slowly move it to the edge of the field of view. If you have to refocus it slightly to get the sharpest image then your 'scope is probably showing some field curvature. Distortion is usually seen when using wide-angle eyepieces on short focal ratio 'scopes. It comes in two flavors; pincushion (positive distortion) and barrel (negative distortion). These distortions are best seen during daylight hours by pointing your 'scope at a flat roof and looking for bending of the image near the edges of the field.

Distortion is very hard to correct completely, and only the best (read most expensive) eyepieces seem to be able to correct for it adequately. The good news, especially if you're a dedicated sky gazer, is that it will have little or no effect on the quality of the night time images your 'scope will throw up and so for the most part can be ignored.

The final aberration featured in the table is the defocus aberration, which you can see scales inversely with the square of focal ratio. This provides a measure of how easy it is to achieve precise focus. In other words, telescopes with greater focal ratios enjoy greater *depth of focus*. Thus an f/8 'scope will be four times $(8/4)^2$ easier to focus than an f/4 'scope of the same optical quality.

Focusing an f/4 Dob can be an exercise in frustration, especially during periods of poor and mediocre seeing, but it is considerably easier in an f/8 Dob under the same conditions. It has been argued elsewhere that depth of focus is strongly linked to image stability (the images will be four times more stable under all conditions of seeing. This author's own research conducted in collaboration with Vladimir Sacek (creator of TelescopeOptics.net) has shown that greater depth of focus is best seen as a tool to achieve best focus position. What's more this research has shown that depth of focus is protective of a telescope's best performance level.

The important general principle to take into account is that when the focal ratio of the mirror is increased all the aberrations fall, so you are more likely to get better images in an f/8 Dob than say an f/4 Dob of the same aperture. That is not to say that high F ratio mirrors are that much easier to make than their low F ratio counterparts. To see why, consider making a 6-inch f/12 spherical mirror and then think about the differences you'd have to introduce to the same mirror to create a parabolic geometry. The difference would be minute! All it takes is just a smidge too much grinding to overshoot the mark! Another problem with high F ratios is that it rapidly increases tube length. Setting up a 10-inch (25-cm) f/5 Dob is a one-person job; but an f/10 'scope of the same aperture is certainly sure to require at least two. That's why the vast majority of commercial Dobs in apertures over 10 inches have very fast focal ratios (f/3.5 to f/5).

How would you rate the image quality in your Dob? OK? Good? Magnificent? A traditional way of measuring optical quality is to specify how well the mirror is figured. Because the differences between a good mirror and bad mirror can be minute, it simply isn't convenient to express errors in everyday units. Instead some opticians prefer to express the error in terms of the fraction of the wavelength of yellow-green light the primary mirror deviates from that of a perfect optic. This color of light has a wavelength of 550 nanometers. One nanometer is one billionth of a meter. An OK mirror will be figured to an accuracy of ¼ of a wave, that is, the microscopic irregularities in the shape of the mirror cannot be more than about 140 nanometers in order for it to operate satisfactorily under most conditions. Such a mirror is said to be *diffraction limited*, which means that the optics are constrained by the wave nature of light itself and not by any flaws in its optical figuring.

Who gave us that idea? That honor goes to the nineteenth-century physicist, Lord Rayleigh, who reckoned that an image distorted by anything more than ¼ wave of yellow-green light would appear *obviously degraded* to the eye. This is called the Rayleigh limit. Of course, it stands to reason that a primary mirror corrected to an accuracy of, say, 1/8 of a wave has an even better figure, but would you notice the difference in the field? Careful observers would definitely say yes. Tests conducted by Peter Ceravolo and published in Sky and Telescope, March 1992 suggest that telescopes with final images of less than ¼ wave P-V wavefront were not as revealing of fine planetary detail as those which were working at 1/8 wave or better. A Newtonian primary mirror that is corrected to an accuracy of ¼ of a wave will show some nice detail on the planets but not nearly as much as an identical reflector corrected to, say, 1/6 or 1/8 of a wave. That said, there is a limit to how much the human eye can discern. In typical tests, most people are not likely to see a difference between an mirror corrected to 1/8 of a wave and one that is corrected to a 1/10 wave accuracy. Dave Bonandrini, a very experienced user of Dobsonians from Ann Arbor, Michigan, in the United States., said this about mirror figure. "I think most astronomers would be fine with a telescope that has a 1/4 wavefront," he said. "If set up side by side, many astronomers can see the difference between a 1/4 and 1/8 wave. I have never seen a mirror test over 1/20 wavefront. Whenever we have tested a 1/30 wave mirror, it usually is well under 1/8 wave."

Surface accuracy is all well and good, but it doesn't tell the whole story. Errors in the figure of the mirror surface making up the objective can lead to increased spherical aberration, coma, distortion, field curvature, and astigmatism (the five Seidel errors). To this end, optical engineers have devised an even better way of expressing optical quality – enter the Strehl ratio.

To understand this quantity better, picture again the image of a tightly focused star seen at high power through the telescope. The star will not be a perfect point but will instead be spread over a tiny disk of light called the Airy disk surrounded, in ideal conditions at least, by of series of diffraction rings. This is what opticians call a diffraction pattern. In 1895, the German mathematician and astronomer Karl Strehl computed what the diffraction pattern of a perfectly corrected lens (or mirror) would look like, with a central peak intensity (representing the Airy disk) surrounded on either side by a series of peaks of progressively less intensity.

A *real* mirror, on the other hand, will have some optical aberrations that will leave their mark on the diffraction pattern observed. For example, a mirror might display some coma and so some of the light never gets focused tightly inside the Airy disk, resulting in a decrease in the peak intensity in its diffraction pattern compared to a perfect lens. Other optical errors, such as spherical aberration and astigmatism, for instance, also leave their mark on the diffraction pattern. And yes – it inevitably reduces the peak intensity of the Airy disk. Put another way, an optically perfect telescope will place about 87 percent of the light inside the Airy disk and the rest is to be found in the surrounding diffraction rings. All real-world telescopes place less than 87 percent of the light inside the Airy disk.

There is a neat way to calculate the Strehl ratio of your 'scope given either your Peak-to-Valley (P-V) error or your Root Mean Square (RMS) error. If ω is the Root Mean Square (RMS) error of your mirror, then its corresponding Strehl is given by:

$$S^{\sim e}-(2\pi\omega)^{2}$$

For example, if we set $\omega = 1/13.4$ RMS wavefront error, then the corresponding Strehl will be 0.8. Such a Strehl value is considered to be diffraction limited. The best-figured mirrors have Strehl ratios considerably higher than this (>0.98 is possible). Finally, it is the author's opinion that under good conditions, where seeing and localised thermal effects do not degrade the image, observers will likely not see much difference between a mirror figured to an accuracy of a smooth ¼ wave and that delivered by one figured to a higher accuracy.

Tube Design

The length of a Newtonian reflector is basically governed by the focal length of the primary mirror. Because the instrument is not a compound design, the effective length of the telescope is very close to the focal length of the primary mirror. Thus, an 8-inch f/6 (20 cm) instrument will have a physical length of at least 1.2 m and a 10-inch (305-mm) f/5 instrument will have a length of the order of 1.5 m. By comparison, compound designs such as Schmidt Cassegrains or Maksutov Cassegrains of equal aperture will have a length less than half that of a Newtonian. The long tube of the Dobsonian (especially closed tube designs) may create some transportation issues. The 10–12 inch (25–30 cm) seems to be the upper limit for solid tubes. That's because a 'scope this size will fit across the back seat of almost any car. Going bigger than this, the advantage is tipped in favor of a segmented design.

Of course, this problem is considerably reduced by making the tube so that it comes apart into two or more sections, creating so-called segmented or truss tube designs. The Newtonian reflector suffers from the fact that whether one uses a solid tube or a truss tube, the optics are left wide open to the effects of dew, air currents, dust, dirt, and whatever else comes along. This is a real disadvantage when compared to the closed system of the refractor, Schmidt Cassegrain or Maksutov design. The advantage of the closed system is that the primary mirror and secondary mirror are kept sealed and remain clean. Added to this is the idea that tube currents cannot form and degrade the final wave-front.

Focusers

If you look through a variety of 'scopes of different focal ratios you'll soon notice a trend. The higher the F ratio, the easier it is to find the point of best focus. A 6-inch (15-cm) f/8 Dob is a breeze to focus; a 6-inch f/5 is considerably trickier. When you get to f/4; a high quality focuser almost becomes a necessity as the depth of focus (which scales directly as the square of focal ratio) is so shallow that the slightest shift can make all the difference between a sharp and a fuzzy image.

The first commercial Dobs to hit the market back in the 1980s and 1990s came equipped with simple but functional rack and pinion focusers. Lubricated by grease, they were always prone to stiffening during cold snaps. Frequent users had to "relube" them from time to time to keep them working at all. These days, things have definitely changed for the better, with even many budget-priced models – the "Econo-Dobs," to use Phil Harrington's phrase – now sporting silky smooth Crayford-style focusers that are definitely a mark up from their rack and pinion counterparts.

That said, some observers have seen the need to upgrade their standard Crayfords, replacing them with high-quality dual speed focusers made by third-party companies. These high-end focusers allow Dob users to attain and hold precise focus much more effectively. In addition, an oversized knob on the top of the focuser body can be tightened to lock it in place for photographic applications. We'll be looking at a few nice focusers that you can retrofit onto your existing instrument in a later chapter.

Mirror Cooling

If you take your Dob out from a warm room to the cool night air and immediately begin to observe, chances are you'll be disappointed by the views it throws up. As soon as you uncap your telescope, the tube begins to fill up with cold ambient air, and the primary mirror starts to acclimate to the cool of the night air. Until it reaches the ambient temperature of the atmosphere around you, the *boundary layer* of warmer air coming off the primary sits just above the mirror surface, causing bad seeing. A quick look at a planet or a star shows it "boiling" in the eyepiece. As you'd expect, the problem gets worse as mirror size increases, and if the temperature continues to drop all night the mirror might never catch up. In that situation, the boundary layer will never disappear.

One excellent solution to this problem is to install a cooling fan behind the primary mirror cell that blows cold air onto the cell, cooling it off to ambient temperature much more quickly while also helping to remove that boundary layer. That said, it still involves a bit of waiting for the mirror (especially for an 8-inch aperture Dobs and larger) to cool to ambient, even with the help of the fan, but nothing like the few hours it takes without one.

The primary advantages of cooling fans are most noticeable when viewing the planets, splitting tight double stars, and other high magnification work. The image is more stable – that is, less prone to degrading – and the images of stars are tighter. In general, you do want to keep the air moving when you are observing; not only will it help the 'scope track a falling temperature, but it will also keep the boundary layer under control. Some observers turn their fans off while actually observing, claiming that it introduces tiny vibrations that can disturb the image.

Mirror cooling can be as simple or as complicated as you want it to be. You can make your own fan for a 6-inch f/8 Dob with a $10 computer fan. It attaches to the rear tube with Velcro and doesn't cause much in the way of vibration issues while observing – even at fairly high powers. If you do see some fan-induced vibrations, you can always experiment by lowering the voltage a little. With a modest set up, a 9 V supply provides effective air flow. Your mileage may vary (Fig. 2.5).

Fig. 2.5. A built-in mirror cooling fan (Image credit: Andy Sheen).

Pause for Thought – The Newtonian Logic

On a cold January day in 1672, the thirty-year old Isaac Newton presented an entirely novel type of telescope to England's Royal Society, where it aroused great interest. He had succeeded in making a mirror with a spherical curvature, slightly less than 1½ inches (3.7 cm) in diameter. The mirror was made of a copper-tin alloy, to which Newton had added a bit of arsenic to make it easier to polish. Above this primary mirror Newton placed a small, flat secondary mirror at a 45-degree angle, to reflect the light into an eyepiece mounted in the side of the telescope tube. And though he was not the first to suggest the use of mirrors in the design of astronomical telescopes, his little reflector changed the world of astronomy forever. Though it was only 15 centimeters in length, the instrument had a magnification of about 40 and could outperform lens-based instruments more than two meters long.

It is now 450 years on, and Newtonian reflectors have improved beyond measure and still command respect from an army of beginning stargazers and seasoned veterans alike. Their great strength is their affordability, embodied in the fruits of the Dobsonian revolution, provide backyard astronomers with arguably the most "bang for the buck" of all telescopes, serving up images that are sharp, detailed, and free from the false color that plagues refractors of the same size. A Newtonian reflector uses a single parabolic mirror to gather light from distant objects. Light enters the tube, traveling down to the mirror, where it is then reflected forward in the tube to a single point called the focal plane. A flat mirror called a "diagonal" intercepts the light and directs it out the side at right angles to the tube through to the eyepiece for easy viewing.

Newtonian reflector telescopes replace heavy lenses with mirrors to collect and focus the light, providing an impressive amount of light-gathering power for the money. You can have focal lengths up to 1,200 mm and still enjoy a telescope that is relatively compact and portable. Let's look at some of the many advantages connected with Newtonian reflectors that make them superior to virtually all other optical systems as serious, all-around performers.

For one thing, they are usually less expensive for any given aperture than comparable quality telescopes of other types. Since light does not pass through the objective (it only bounces off a mirrored surface), exotic glasses are not needed; the material only needs to be able to hold it to an accurate figure. Because there is only one surface that needs to be figured (as opposed to four in a refractor) it is easier for amateur telescope makers (ATMs) to fashion their own objective. A short focal ratio can be more easily obtained, leading to wider field of view. Long focal length Newtonian telescopes can give excellent planetary views.

Another great utility afforded to Newtonian reflectors is that they can be made in large enough sizes to satisfy the fundamental requirements for the general observer without experiencing extremely high costs. For example, an average 12-inch (250-mm) f/5 Newtonian can be made to sit comfortably within the cradle of a simple Lazy Suzan-style alt-az mount and can be set up for observing within minutes (if properly acclimated).

Such an instrument, once cooled down to ambient temperatures, can detect objects some nine times fainter and can resolve details three times finer than the best 4-inch (10-cm) refractor that usually costs two or three times as much. And although they are certainly not grab 'n go instruments, large Dobsonians – especially those of the truss-tube variety – can be transported long distances in the back of a small car and can be set up within minutes.

In addition, the simplicity of the Dobsonian mount means that very large backyard 'scopes can be acquired for a relatively small financial outlay. This author once had the opportunity to look through a massive 30-inch (75-cm) monster at a star party, and the views were spectacular to say the least, though it took a while to get used to maneuvering the ladder to reach the eyepiece. We spied the Ring Nebula (M57). What were the views like? Well, words such like "spectacular," "huge," "colorful," and "compelling" spring to mind. Of course, Dobsonian-mounted Newtonians have also entered the electronic age. Several manufacturers now sell encoders, for both the right ascension and declination axes, that allow you to quickly locate thousands of galaxies, star clusters, and nebula simply by pushing the 'scope to a particular spot in the sky. What's more, large Dobsonians can now be placed on specially designed, tilted platforms that enable the 'scope to track objects automatically as they move across the sky. And most recently of all, alt-az mounted Dobs have now been empowered with full GoTo capability.

So why doesn't everyone use a large aperture Dob instead of coveting smaller refractors or ultra-compact catadioptric designs? The answer is complex and varied. Some diehards can never get excited about Dobsonians, or even Newtonians in general. Newtonians have issues – there's little doubt about it – that arise both from their complex nature relative to simpler designs (such as refractors), as well as thermal mismanagement. For others, the problem lies in the quality of the mirror. Although they don't suffer from chromatic aberrations, they frequently suffer from miscollimation issues, which creates the wrong impression if the owner isn't careful enough to check the alignment of the optics on a regular basis. Other mirrors are just poorly designed and throw up astigmatism, coma, and modest amounts of spherical aberration. As we've seen, the shorter the focal length of the mirror, the more difficult these aberrations are to control, with the result that top-quality, "fast" mirrors are more expensive to buy, and focal ratios much below f/3 or f/4 are extremely difficult to make with any accuracy. That said, a long focal length (f/8 or slower) Newtonian is hard to beat as a lunar and planetary 'scope, especially in apertures from 6 to 10 inches. because these aberrations are very much reduced.

The Newtonian reflector requires a smaller secondary obstruction than any other reflecting telescope design. Compound reflectors of the Cassegrain and Gregorian types, either in classical or Schmidt Cassegrain or Maksutov configurations, require secondary obstructions having a physical size of at least 25–35 percent of the diameter of the primary mirror. Newtonian reflectors, by contrast, even of f/6 design, can easily accommodate secondary mirrors less than 20 percent of the diameter of the primary mirror and still give full illumination to a reasonable field of view. Better still, for focal ratios of f/8 and higher, secondary obstructions of 15 percent and smaller can be attained. No other reflecting optical system can do this without resulting in extreme proportions.

The impact of the secondary obstruction on observing is most readily noticed in attempting to resolve fine, low-contrast planetary detail. Telescopes having central obstructions in excess of 25 percent begin to degrade the image noticeably in the arena of low contrast resolution. When the size of the secondary reaches 30–35 percent, as it does in many commercial Schmidt Cassegrain telescopes, the reduced ability to resolve fine detail becomes extremely noticeable. This obstruction and the so-called diffraction spikes caused by the support structure (called the spider) of the secondary mirror reduces contrast. Visually, these effects can be reduced by using a two- or three-legged curved spider. This reduces the diffraction intensities by a factor of about four and helps to improve image contrast, with the potential penalty that circular spiders are more prone to wind-induced vibration. Although a four-legged spider causes less diffraction than a three-legged curved spider, the latter often gives a more aesthetically pleasing view.

So there you have it: Dobsonians are serious instruments for lunar, planetary, and deep sky observing. Although not as readily portable as a small refractor or Schmidt Cassegrain or Maksutov instrument, such an instrument will optically match or outperform all other forms of astronomical telescopes inch for inch of aperture in larger sizes. That said, a Newtonian reflector requires slightly more care and consideration in use, but will be significantly less expensive to construct than any of the other telescope types. The point to emphasize here is that the Newtonian reflector is in no way a substandard instrument when compared to other compound reflecting optical systems or refractors. If the gremlins associated with accurate collimation and thermal management are sorted out, then it is every inch the equal of these instruments and, in some ways, superior. If the instrument is designed well and constructed out of quality materials, the views it will serve up will absolutely amaze you.

You will note that we understand the term "Dobsonian" to mean any simply mounted (usually alt-az) reflecting telescope. It need not be a Lazy

Suzan mount, and it may not even be a Newtonian reflector, but it's got to be easy and intuitive to operate. John Dobson himself would undoubtedly be happy with this general description.

All that said, we're now ready to explore the rich milieu of the Dobsonian telescope. In the following chapters, we'll explore the ways in which the Dobsonian revolution has developed over the last decade into an astonishing array of instruments, from tiny hand-held rich-field 'scopes to towering giants that would hold their own or even outperform many observatory class telescopes. To begin with, let's take a look at the most diminutive of the commercial Dobs, the new breed of starter 'scopes that get so many youngsters and older beginners hooked on sky gazing.

CHAPTER THREE

The Mini-Dobs

What's the ideal starter 'scope? Well, it should have decent optics, be portable enough to set up in just a few minutes, and have enough light grasp to bag many of the brighter treasures of the night sky.

Many claim a small refractor fits the bill. But there are compelling reasons to believe that this niche is best filled by a small Dobsonian. In recent years, there has been a great proliferation of small high-quality Dobs that give a lot of bang for your buck. Indeed, a 6-inch (15-cm) f/8 Dob is a very decent performer and could occupy a curious mind for a lifetime if properly looked after.

To start the chapter, let's call your attention to smaller instruments that have appeared in time to celebrate the 2009 International Year of Astronomy. Tiniest of all is a 3-inch 'scope called the Celestron FirstScope. This demure Dob has a focal length of 300 mm (focal ratio f/3.95). So what do you receive for your money? Well, you get the 'scope and two eyepieces, giving powers of 15× and 75× magnification, which will serve up decent views of the Moon, pretty doubles such as Albireo, Mizar, and Alcor, and a suite of bright deep sky objects such as the Pleaides (M17) star cluster or the majestic Orion Nebula (M42) from your light polluted backyard. With one of these to entertain the guests, you'll be sure to give them an evening they'll not forget in a hurry – weather permitting, of course (Fig. 3.1)!

N. English, *Choosing and Using a Dobsonian Telescope*, Patrick Moore's Practical Astronomy Series 1, DOI 10.1007/978-1-4419-8786-0_3,

Fig. 3.1. The diminutive Celestron FirstScope (Image credit: OPT).

As you can imagine, the FirstScope is very portable, and as far as storage is concerned, it's a no brainer! Just leave it in plain sight (with caps on), until you're ready to take it out of doors. So it is inexpensive, comes assembled, can be stored anywhere, and looks cute. But what's it got under the hood? The ability to grab'n'go, no fancy set up time required – and hey, that makes a big difference. It can be unpacked, assembled, and outside observing within 3 min. Also available for an additional $20 is an accessory kit – the Celestron FirstScope Accessory Kit, which comes with two more eyepieces, giving you 24× and 50×; a Moon filter, which allows for better imaging and viewing during a full Moon; a carrying case for your 'scope; a 5×24 finder 'scope; and a CD ROM, filling you in on the basics of astronomy, familiarizing you with stars, planets, and constellations and giving you the ability to print off star charts. Very cool, definitely five out of five stars. The 'scope retails for just $39.95 and comes with a two-year warranty.

Not to be outdone, Sky Watcher also launched its own version of this mini Dob – the 76 Heritage 'scope. Like the Celestron FirstScope, the 76 Heritage has a tiny 3-inch mirror with a 300-mm (f/4) focal length. Weighing in at 1.75 kilos, it comes with two eyepieces (a 25 mm and 10 mm, yielding 12× and 30×); it sits on a cute wooden alt-azimuth mount. The Sky Watcher Heritage 76 retails for about $10 more than the Celestron version. So it is worth the higher cost? (Fig. 3.2)

The Sky Watcher is perhaps more solidly made and comes with an attachable finder 'scope. Optically, however, there is little to choose between them. They both have tiny spherical mirrors and so are clearly

Fig. 3.2. The Sky Watcher Heritage 76 'scope (Image credit: Harrison Telescopes).

aimed at the young novice observer or just the plain curious. Though they are fine instruments as far as they go, they are hard to recommend to anyone with a serious urge to begin the hobby. For that, you'd be much better served with a 4.5 inch (114-mm) or better still a 130-mm (5.1-inch) Dobsonian.

Orion Telescope & Binocular has also launched its vision for the mini-Dob in the form of the SkyScanner 100-mm table top reflector ($99). This little telescope, finished in a nice maroon colored aluminum tube, features a fast f/4 parabolic mirror and is just the right size to sit on your desk. The 'scope and mount also come with two 1.25 eyepieces yielding 20× and 40×, and an EZ finder II reflex sight. Owners report fairly decent views up to powers of about 100×, but at f/4 you'll definitely need well corrected (read more expensive) eyepieces to get the best wide field performance out of this telescope (Fig. 3.3).

Another nice feature is that this telescope cools down quickly after being taken outside . Within 20 min or so it can be serving up decent images of some of your favorite star clusters, such as M35 in Gemini,

Fig. 3.3. A nifty performer: the Orion SkyScanner 100 (Image credit: OPT).

or the Coathanger asterism on the border between the constellations of Vulpecula and Sagitta.

Replacing the stock eyepieces with some higher quality units will allow you to see a lovely crescent Moon at 75× with a wealth of high resolution detail popping into view. This is no toy telescope. Saturn's rings are also a sight for sore eyes, but when you crank up the magnification to 150× using a 2× Barlow lens, the image can be a tad too soft. This unit at least, displayed small amounts of spherical aberration and a trace of astigmatism, too. Though there is no provision to accept 2-inch eyepieces, it's probably a blessing in disguise. The amount of coma you would get with such an ultra-low power, wide angle eyepiece might put you off observing. All said, it's one sweet package for its modest price tag.

The 4.5-inch Brigade

Quite a few sky watchers made their debut with small reflectors, typically with a primary mirror diameter of 4.5 inches (114 mm). Even a while back, they offered a significant step up in light-gathering power and resolution, though it can be hard to get used to peering 'down' into the eyepiece while looking at objects perched high in the sky. Today, this size telescope is still proving popular among novice astronomers and more experienced ones to boot. One of the best to hit the market in recent years is the Orion Starblast 4.5. Launched back in 2003, this good-looking Dob combines decent optics in an ultraportable package. You'll notice the difference in high power views immediately when you look through this telescope. That's due to the presence of a good parabolic mirror at its heart.

Easy to take out for a few minutes of viewing, the Starblast 4.5 is a surprisingly good performer on just about any target. The unit often comes with perfect collimation right out of the box, and the mirror is center spotted for easy collimation. Both the secondary and the primary mirrors can be easily adjusted, and the main 'scope also comes with a collimation tool. The Starblast 4.5 is also supplied with an EZ red dot finder, which makes aiming the 'scope easy. The altitude tension can be adjusted easily and azimuth is smooth. The focuser is a nice rack and pinion, but some users have found that it can be a bit stiff. Other nice touches include an eyepiece rack that holds three occulars. The telescope also comes complete with a pair of eyepieces of 17-mm and 6-mm focal length of decent quality. The 17 mm does a great job framing objects within its 2 degree field of view. The 6 mm gives impressive medium power views at 75× of the Moon and the bright planets, though not with very good eye relief. You can frame the Andromeda Galaxy (M31) with a 24-mm wide angle eyepiece, its bright nucleus decorated on either side by bright spiral arms stretching fully 3 degrees across.

This telescope is no slouch either when it comes to soaking up high magnification. During some exceptionally steady winter nights you can observe Mars and Saturn at powers up to 200×. Though the images might be somewhat soft, Cassini's division is easy, as is shading on the planet surface.

For double star observing, you could use it at 180× on most good nights. Indeed, all in all, the Starblast 4.5 performed far better than expected from its modest price tag. It is a very solidly recommended entry level 'scope for anyone who wants decent views on a limited budget or in an ultraportable package (Figs. 3.4 and 3.5).

Fig. 3.4. The Orion Starblast 4.5 (Image credit: OPT).

Fig. 3.5. The Sky Watcher Heritage 130P.

While the Sky Watcher 76 is a good telescope, most novices will soon run out of things to see with it in due course. More aperture and better optics will open up a whole suite of new wonders that are quite inaccessible to a 3-inch 'scope of any size. Aperture fever will soon set in, and when it does, there's a palliative of sorts. Enter the Sky Watcher Heritage 130P.

This 5.2-inch (130-mm) f/5 Dob also boasts a well figured parabolic mirror, like the Star Blast 4.5, only with 30 percent greater light-gathering power. That said, the larger aperture Sky Watcher telescope actually weighs in at just about the same weight (6.2 kilos) as the Orion 'scope. That's due to the presence of the flex tube design. The upper part of the tube is essentially "missing" and collapses down onto the bottom end of the tube for easy storage. This "flex tube" design actually works remarkably well, and it holds collimation even after repeated outings to the garden (Fig. 3.6).

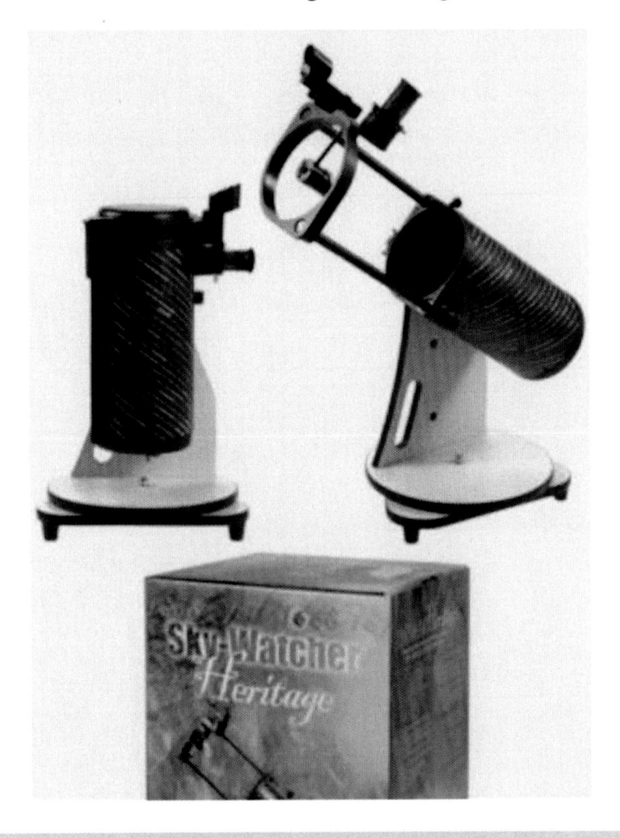

Fig. 3.6. The Sky Watcher Heritage 130P (Image credit: OPT).

Because the 130P is only 70 cm long when the tube is fully extended, it is necessary to mount it on a sturdy garden table. The all-wood Lazy Susan mount moves smoothly in both azimuth and altitude, and the supplied red dot finder makes aiming at celestial objects a breeze. Indeed, the collimation is almost dead on out of the box. While the two supplied eyepieces (budget Plossls of 25 mm and 10 mm focal lengths yielding 26× and 65×, respectively) are adequate performers, this 'scope will certainly benefit from using higher quality eyepieces. Showpiece objects such as the first quarter Moon and the Orion nebula are a joy to behold, and the 'scope takes magnification fairly well. Certainly 180× is not beyond its capabilities on a good night. It can provide crisp views of Saturn near opposition using a quality 4-mm eyepiece. Star testing at the same high powers revealed a well figured mirror with a touch of under-correction, but the optics in the sample tested appeared to smoothly figured. Conclusion? It is a fine beginner 'scope, possibly the best one to be had under $200 on the market.

The 6-inch Crew

When you make the move to a 6-inch (15-cm) aperture Dob, the sky and its showpieces really start to reveal their finer details. During the 1960s and much of the 1970s, a 6-inch reflector (remember the celebrated RV-6?) was thought to be the best choice for a well-heeled amateur astronomer, and the same is largely true today. The 6-inch Dobs are very popular indeed and come in a variety of forms to suit the differing needs of observers. They are highly portable – even in the longer focal lengths – and deliver views of the night sky that would keep even an experienced observer happy for years. Indeed, there is a very strong case for believing that a well-made 6-inch Dobsonian is the very best starter 'scope money can buy for a serious beginner. Such a 'scope has the potential to outperform even premium 4-inch (10-cm) refractors costing many times more under ideal conditions.

Discovery Telescopes, based at Mission Viejo, California, is a trusted name in manufacturing high quality optics for Dobsonian enthusiasts. Although it makes a range of larger instruments, the company also sells a very handsomely finished 6-inch f/8 DHQ Dob ($400) for the amateur astronomer. Tom McCarthy from North Carolina described his experience with this well made 6-inch f/8 Dob. "I am very happy with this instrument," he said. "The optics are very good, giving very crisp, clear views of Moon and planets. On a night of average seeing, I can see at least three bands on Jupiter and several more on the best nights. The 'scope takes magnification well. Saturn and its Cassini division are simply gorgeous, and contrast is easily good enough to allow banding on the planet's globe to be made out. The Orion nebula (M42) is beautifully presented as a seething mass of incandescent gas. My Discovery 6 DHQ shows the stars to be pinpoint sharp and displays their true colors faithfully. When I took it to a local star party this little Dob could definitely hold its own with any 'scope out there of comparable aperture, even telescopes there that were much more expensive. This Dob is a real bargain, and for me at least, it's a keeper. Sadly, this very fetching small Dob is no longer produced by Discovery Telescopes, but many remain on the used market and can be had for relatively little money."

The Starblast 6

Thankfully, there are many other Dobs of similar aperture to choose from. One great performer in this aperture class is the Orion Starblast 6, made in the same user-friendly style as its smaller sibling, the Strablast 4.5, discussed previously. Scott Holland, from Lowell, North Carolina near Metro Charlotte provided his thoughts on this telescope, having owned and used it for quite some time:

"I have moderate light pollution at my home in Lowell, but as a member of the Charlotte Amateur Astronomy Club, I have access to a very nice dark sky site some 60 miles south in South Carolina. Over the 40+ years that I have enjoyed amateur astronomy, I have used telescopes that range in size from my first, a 30-mm spyglass from the Sears & Roebuck catalog, a 10-inch SCT computerized go-to 'scope, and I've even been privileged to observe through the 23 inch Alvan Clark "War of the Worlds" refractor now located at Charles E. Daniel Observatory near Greenville, South Carolina. When I first saw the advertisement for the Orion 4.5-inch Star Blast telescope, I shrugged it off as just a kid's toy. After all, a 4.5-inch f/8 'scope gives respectable images, but an f/4 focal ratio demands both an excellent parabolic figure and careful collimation, or images would be mush. For the price, I doubted that either was true. Boy, was I wrong! One night, a couple of years ago, I loaded up my old Celestron C8 Ultima, a bunch of eyepieces, charts, and a red lamp and headed for our dark sky site. When I arrived, a few other club members were already there, and one of our guys had just set up his new Orion Star Blast 4.5 on a modified bar stool. There was also a 10-inch Dobsonian reflector and a 6-inch Maksutov set up on the club observing pad. The seeing and transparency were very good that night. Before the night was over, we all were surprised and impressed by the little Star Blast's performance. If it only had a bit more aperture!

"Someone else must have been thinking the same thing, because it wasn't long before ads began to appear for the new Orion Star Blast 6, a 150-mm f/5 tube on a larger, but very similar, style mount for under $300.

"Screwdrivers, wrenches, and pliers are usually needed to assemble the mount. Here, the mount was pre-assembled. All you needed to do was attach the tube rings to the altitude hub, open the rings, and install the optical tube. Next, the eyepiece holder was mounted by sliding the bracket down over the screws on the braces, then tightening the screws. It couldn't be easier (Fig. 3.7).

"The location of the eyepiece holder also seems to provide some additional bracing of the altitude axis. The vertical braces have cutouts to be used as carrying handles, but the edges of the laminate in the cutout area can be rough on the hands if you're not careful. The altitude axis also has a Rosette knob for tension adjustment located between the vertical braces. All one need do is match this tension setting to the preset azimuth tension for smooth movement and tracking (Fig. 3.8).

Fig. 3.7. The adjustable friction knob on the Starblast 6 mount (Image credit: Scott Holland).

Fig. 3.8. The Orion 'scope comes with two good quality Plossl eyepieces (Image credit: Scott Holland).

Fig. 3.9. The primary mirror cell has three spring loaded collimation knobs and three locking screws (Image credit: Scott Holland).

"The optical tube is a standard Orion 6 inch with an f/5 primary, adjustable mirror cell, adjustable secondary, 1.25-inch rack and pinion focuser, a Vixen-style finder shoe, and an Orion EZ Finder II Red Dot. Orion sells this tube on several different mounts, including the Versa-go, and on an equatorial mount (Fig. 3.9).

"The 1.25-inch focuser has a lot of plastic in its design, but it does a decent job with the f/5 primary. The EZ Finder II Red Dot is a great finder for use with a short focus, rich-field Newtonian, especially if you're young. My only problem is bending down at the back of this short telescope to get a view through the finder to get the 'scope aimed. For an older person, you should consider substituting a RACI finder in the shoe and you're good to go. Orion provides two very good Sirius Plossl eyepieces with the 'scope, a 25 mm and a 10 mm for 30× and 75×, respectively. They also provide a collimation cap for rough collimation of the mirrors (Figs. 3.10 and 3.11).

"As can be seen in the photo (Fig. 3.11), the primary is center marked for ease of collimation. Use a sight tube and a laser collimator to align the secondary. Then use a Cheshire eyepiece to adjust the primary. Do fine adjustment by defocusing on a star at high magnification.

Fig. 3.10. The secondary is held in a fully adjustable holder by a four-vane spider (Image credit: Scott Holland).

Fig. 3.11. The secondary is held in a fully adjustable holder by a four-vane spider (Image credit: Scott Holland).

Overall, the Star Blast 6 can be said to be an excellent starter 'scope and a very good value. From star testing, it seems that the optics in this 'scope are better than ¼ wave and probably more like 1/6th wave. It does great on wide field views of clusters and does a respectable job on the Moon

Fig. 3.12. Ready for action: the Orion Starblast resting on a garden table (Image credit: Scott Holland).

and planets. On the negative side, it is short. You might want to put yours on a table for easy access, but with the table, vibration is a problem for a time after moving the OTA. A better, more stable table would probably help (Fig. 3.12).

"One final note. Orion has recently released the new version of the Star Blast 6, called the Star Blast 6i. This version sells for about $400, has encoders in the altitude and azimuth axis, and comes with the Orion IntelliScope feature. That makes the 6i a "push-to" 'scope with digital setting circles. The computer is loaded with 14,000 objects to make locating "faint fuzzies" an easier task. However, just remember that the Star Blast 6 and 6i are only 150 mm telescopes, so a dark sky site would be necessary to take full advantage of the giant database. Man, we've come a long way in 40 years!"

Actually, Scott's account is fairly typical of what other amateurs have said about the Orion Starblast 6 Dob, and considering you can have a computerized object locator for just a little bit more money, it's hard to see why anyone wouldn't consider it one heck of a bargain in today's market.

Going Long

As the discussion on Newtonian optics pointed out in Chap. 2, there are obvious advantages to having a long native focal ratio. All the aberrations fall off rapidly as F ratio increases, and so, all things being equal, it is easier to get high quality optics in a higher F ratio 'scope than in a lower F ratio. All you sacrifice is portability and field of view. Another plus for high F ratio Dobs is that they are far more forgiving on eyepieces, so even inexpensive ones up their game. For these reasons, long focus Dobs can be very effective lunar, planetary, and double star 'scopes, and some of the best models offer real alternatives to high end refractors costing many times more.

Dave Rogers from Bolton related his experiences with the Orion XT6, a 6-inch f/8 closed-tube Dob. "It is unbelievable to me that a 'scope this good can be had for this price," he said. "With 6 inches you get decent aperture, and the views through this 'scope are very satisfying indeed. You'll enjoy it more because the images are good enough to linger over. Although you can purchase this 'scope with an object locator (the XT6i), I take pride in finding observing targets on my own, so I have no need of the upgrade (Fig. 3.13).

"The 'scope was in nearly perfect collimation right out of the box, and it is mechanically quite adequate. It took me about an hour to put it together. Then out it went into the cool night air. Star tests show that the mirror is quite good. Nice, concentric patterns inside and outside focus with no sign of rough zones or turned edges. Luna turns on such a show with this 'scope that you can easily get stuck there for a while. Jupiter's cloud belts are a breeze, and contrast is very good. The telescope takes magnification well on good nights. I've had it at 300× on Saturn, and the image showed no signs of breaking down. Globular clusters like M13 look like diamonds scattered on black velvet. Open clusters like M37 and M38 in Aurigae sparkle at medium power, and fainter galaxies like M81 and M82 are clearly discerned. Bright nebulae will seduce your eyeballs

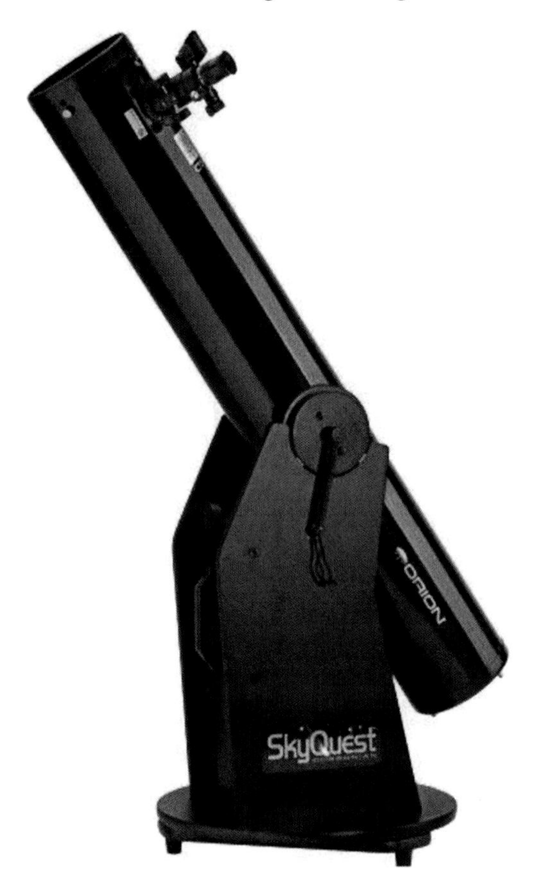

Fig. 3.13. The Orion SkyQuest XT6 Dob (Image courtesy: OPT).

until you realize that it's well past midnight and you're frozen half to death. At 35 lbs assembled, this is about as much 'scope as I want to be moving about. Many novices think the best telescope is the biggest one you can get your hands on, until they realize that they have bitten off more than they can chew. Unless you are lucky enough to have an observatory, you have to think about the weight, first and foremost. If you decide it would keep you from using this 'scope very often, by all means get a smaller one. Orion's XT4.5 is this thing's little brother; it weighs half as much, and it has gotten uniformly good reviews just about everywhere."

The optics for these telescopes invariably originate from Long Perng in China and for inexpensive 6-inch Dobs they seem to offer terrific value for money, are easy to transport, and deliver the readies when it matters. Chinese optics have come a long way.

Sky Watcher Magic

Even a cursory survey of the amateur market will reveal that a 6-inch f/8 Dob is a popular choice among newcomers and more seasoned observers alike. In the UK, the Sky Watcher Skyliner 150P has proved especially popular. Unlike the other Flextube Dobs sold by the company, the Skyliner Ps reserve the traditional closed tube design. A new one can be gotten for the princely sum of £180 ($279), and it arrives in two boxes. The first box contained the 'scope, finder, and eyepieces. The second box contains the flat-packed Dob mount as well as all the tools. Using the instructions – which are quite legible – you can assemble the entire telescope in a little over an hour (Fig. 3.14).

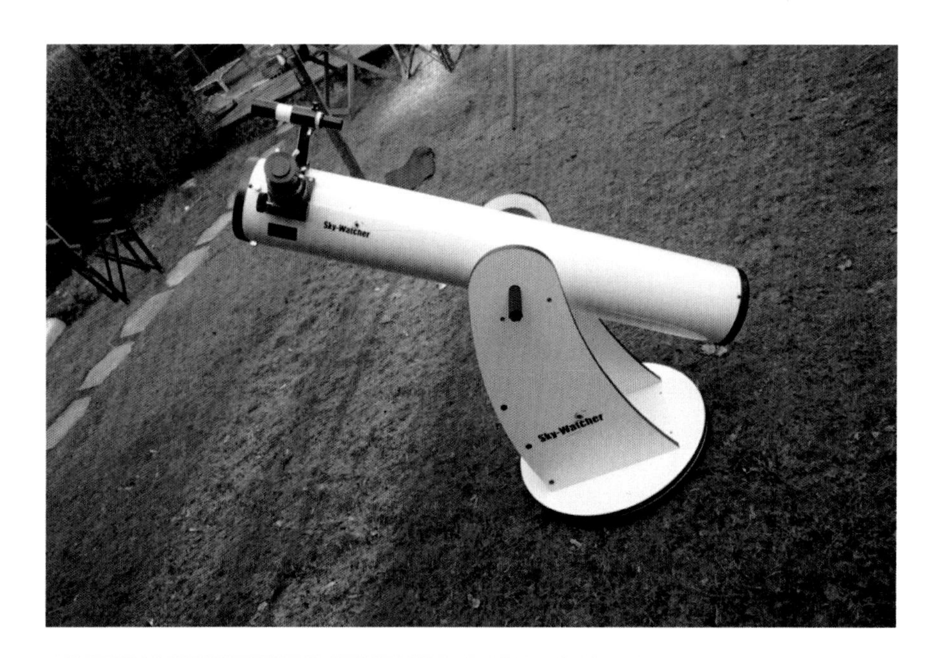

Fig. 3.14. Sky Watcher's Skyliner 150P (Image by the author).

Fig. 3.15. The simple but effective rack and pinion focuser on the Skyliner 150P (Image by the author).

The Skyliner 150P is a no frills 'scope and is very well made for the money. The sturdy, yet lightweight rolled steel tube is finished in white. The lockable focuser is of a simple rack and pinion design lubricated by the well-derided industrial grease. It's a bit stiff, but with its generous focal ratio, achieving best focus is still fairly easy. A quick look through a homemade collimation eyepiece reveals the optics to be well aligned. All of the 1.25 eyepieces tested reach focus thanks to a generous 60 mm of travel (Fig. 3.15).

The 6×30 finder, though cheap and cheerful, makes finding things easy. The whole telescope moves smoothly in azimuth due to a well-made mount with Teflon bearings. Motions in altitude are less smooth, but you can still track objects manually at 200× by getting to know its quirkiness. The primary mirror is centered and marked for precise collimation. The 'scope delivers nice views right out of the box. A star test shows an almost perfectly collimated optic. This is generally true of other Sky Watcher 'scopes. First impressions last. Imagine how disappointed a newcomer to the hobby would be if the mirrors were way out of alignment, and a quick look at the Moon threw up a grotesque blur. Such an experience is potentially capable of scaring a novice right out of the hobby. Sky Watcher seems to be conscious of this potential pitfall and are making positive steps to avoid the problem (Figs. 3.16 and 3.17).

Fig. 3.16. Nice touches, like the carry handle and eyepiece caddy, make both moving the 'scope and observing with it a breeze (Image by the author).

Only a few decades ago, the choice for the serious amateur on a budget was a 6-inch (15-cm) reflector or a 4-inch (10-cm) refractor, and, even today, these telescopes are still broadly accepted as about the minimum required (within their respective genres) to make serious inroads towards the starry heavens. But which of the two – the refractor or the reflector – will show you more? Is one more versatile than the other? Let's answer this question by putting two inexpensive telescopes; a Russian Tal 100R refractor and a Sky Watcher 150P Dobsonian, through their paces over a period of several weeks from a rural backyard.

Fig. 3.17. Mirror versus lens: a shootout between a 10-cm f/10 refractor and a 15-cm f/8 Dob (Image by the author).

Meet the Contestants

The refractor was a Russian-made, late 1990s 4-inch f/10 Fraunhofer doublet. Its descendant, the Tal 100RS, retails for about £270 in the UK for the optical tube (the older Tal 100R was significantly less). The Sky Watcher 150P (£180 in the UK) was a classic Newtonian reflector, with an aperture of 6 inches (15 cm) and a focal length of 1,200 mm (f/8). It had a 1.25-inch secondary (minor axis), which gave it a central obstruction of 21 percent by aperture. That said, the theoretical superiority of the reflector at the outset should be noted. After all, it had 50 percent greater resolution and collects the same amount of light as a 5.5-inch (140-mm) unobstructed 'scope. The instrument sits in the Lazy Susan cradle of a classic Dobsonian mount. The Tal was placed on a sturdy, ash wood TeleVue Gibraltar.

Both 'scopes had a simple rack and pinion design – something that is altogether adequate for 'scopes that possess such generous depth of focus. Well corrected, long eye-relief eyepieces were used, giving roughly equivalent, low, medium, and high magnification views according to seeing conditions. The chosen targets included the Moon, open and globular clusters, a post-opposition Saturn, and a variety of challenging double stars.

The Sky Watcher Dob took about an hour to thermally equilibrate after being taken out into the cold night air from a cool, indoor basement. When an eyepiece was inserted, the 'scope became a bit top-heavy and tended to tip over if the sidebars were not tightened sufficiently. More careful star testing revealed that this 'scope had a good mirror – a nicely figured paraboloid – with concentric diffraction rings seen either side of focus. No signs of surface roughness were noted. If anything, the mirror revealed a tiny amount of under correction, but too insignificant to affect the in-focus images.

Inserting a 24-mm eyepiece yielding 50× and a 1.4 degree field, you would be treated to gorgeous views of the Beehive (M44) and the Perseus double cluster. Stars were presented in their natural hues, with several red giants showing up in the dark space between the main clusters. Stars remained pin sharp, almost all the way out to the edge of the field. Only at the edge of field could you detect very mild amounts of coma and astigmatism.

Star testing the Tal showed well collimated, concentric rings inside and outside focus. Coma and astigmatism were virtually non-existent. Bright first and second magnitude stars threw up purplish haloes at high powers around tight Airy discs – the unavoidable consequence of its achromatic nature. That said, for the majority of objects studied, the secondary color (false color) was unobtrusive. Comparing the views with the Tal refractor (42×), one thing was immediately obvious. The unobstructed optics of the refractor served up noticeably more contrasted views between the stars of these open clusters and the background sky, although it did not pick up nearly so many stars compared with the larger aperture reflector.

M13 provided a powerful reality check. This fifth magnitude fuzz ball in the keystone of Hercules was transformed into a veritable swarm of stellar "bees" in both 'scopes. Comparing the view at 180× in these instruments, the 6-inch Dob was the clear winner. Despite its lower contrast, the reflector's superior resolution and light-gathering power provided a knock-out blow to the aperture-challenged refractor. Whereas only the outer parts of the cluster were resolved into stars in the 4-inch glass, the 6-inch Dob resolved many more stars almost all the way to the center.

High resolution work: Observing the Moon, planets, and double stars requires good optics and steady skies in equal measure. Observing Saturn, both 'scopes served up razor-sharp views of the Ringed Planet at 200×, with hints of the Cassini division visible at the ansae. The Russian achromat imparted the planet's globe with a distinct yellowish hue compared with the straw colored image of the perfectly achromatic reflector. Although more detail could be discerned with the Dob in moments of good seeing, the refractor simply enjoyed more of those moments. Its smaller aperture and closed tube design both worked to its advantage here. On the other hand, the open tube design of the Dob makes it much more susceptible to thermals. The same was largely true when comparing the views through both 'scopes on the Moon. The 6-inch served up delicious low and medium power views displaying a wealth of detail to the eye. The view through the achromat was noticeably dimmer, especially when higher powers were employed. The achromat also showed up some minor false color around crater rims, which cut definition down a notch in comparison to the Dob. But even with a mirror fan (acquired cheaply and affixed to the back of the primary cell with Velcro), the seeing often limited the resolving power of the 6-inch Dob on the resolution of the smallest lunar craterlets and rilles. The refractor resolved to its practical limit more frequently.

Double star observations: A good, long focus refractor has often been cited as the best instrument for resolving tight double stars. This time honored recommendation was put to the test on a variety of challenging pairs for small telescopes. In particular, studies concentrated on three systems – Delta Cygni (mag 2.9/6.3, sep 2.4 inches), Iota Leonis (mag 4.1/6.7, sep 1.9 inches), and Theta Aurigae (mag 2.7/7.2, sep 3.5 inches) – chosen for the fairly close separation and great brightness difference between the primary and secondary stars. The Tal could resolve these systems on average to good nights, but they quite often proved tricky for the 6 inch reflector. Despite its theoretical ability to resolve subarc second separations (0.8 inches, the Dawes limit), contrast-robbing diffraction effects, an open tube design, and larger aperture might all have played their part in reducing the efficacy of the reflector here. That said, on those infrequent nights of excellent seeing, the 6-inch Dob easily outperformed the Tal on all these targets.

So, which 'scope serves you better? Both instruments could be recommended without any hesitation. The 6-inch reflector can outperform a 4-inch glass on nearly everything on good to excellent nights, but the refractor allows you to get decent observations more frequently. So, it seems to all boil down to what you like to observe and how often

Fig. 3.18. The central obstruction due to secondary is a respectable 31 mm (1.25 inches).

you observe. The greater frequency with which the refractor serves up diffraction-limited views could be said to confer greater efficiency, or observational "mileage," to your program despite its smaller aperture and light-gathering power. This is particularly evident with double star observing. As for the reflector; it shows you more, but not nearly as often (Fig. 3.18).

A British-Made Classic

If you're after a better designed classic in this aperture class, then why not consider Orion Optics UK 6-inch (15 cm) f/8 Dobsonian (£399)? The mounting is quite remarkable in and of itself. For one thing, it's made from very strong aluminum plate coated with a hard polymer for scratch resistance. One of the great virtues of this product is the ability of the mount to accept other telescopes. Another notable feature is the built-in capacity to precisely balance the tube, which prevents it from moving when a heavy eyepiece or camera is attached – sadly, an all too familiar occurrence with many other Dobs. The totally re-designed mount gets to grips with a common problem encountered by Dob users using standard Lazy Susans, which usually involves increasing the drag on the declination axis thereby increasing the friction on those bearings and, as a consequence, creating a severe problem in not allowing smooth adjustment of the telescope's position. You'll have none of that with Orion UK's elegant mount. The company even produces an optional variable friction brake that makes even high power observing a comfortable reality. All in all, this 'scope is an exceptionally well engineered piece of kit. But, mechanics aside, how well does it perform? (Fig. 3.19).

Well, the standard model gets you a nicely figured ¼ wave primary with standard coatings. Weighing in at about 29 pounds (13 kg) when fully assembled, this 'scope is very portable. The finder is a bit small,

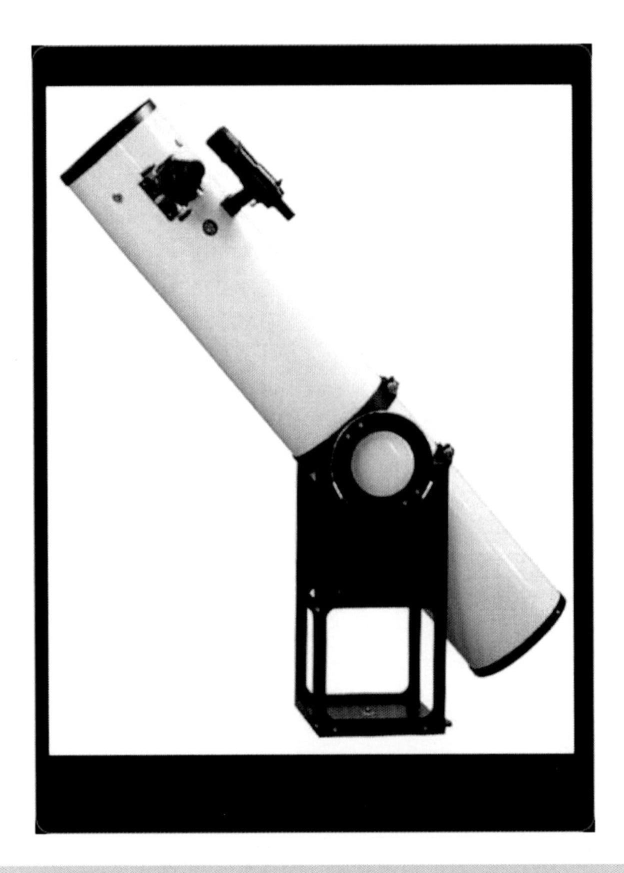

Fig. 3.19. Orion Optics Classic 6-inch f/8 Dob (Image credit: Orion Optics UK).

just a standard 6 × 30 mm, but it does a fairly good job locating most of the brighter objects you'll want to visit with this 6-inch Dob. The Orion UK 'scope is also supplied with a nice 25-mm Plossl eyepiece, yielding a respectful 48× and a near one degree field.

The deluxe model of this telescope comes with a more accurately figured mirror (1/6–1/10 wave or less) and with enhanced (Hi-Lux) coatings, but at extra cost (up to £100 extra). Compared to other mass market instruments, this is an expensive telescope, but you do get what you pay for. If you elect to purchase the standard model, one can always contact Orion to arrange an upgrade at a later stage. Whichever way you look at it, these are nice Dobs and will be sure to satisfy even the most discriminating of observers.

That completes our survey of the small Dob market. Although 6-inch instruments will certainly satisfy a portion of observers, many (if not most) advanced observers crave more aperture and resolution. In the next chapter, we'll be exploring arguably the most popular aperture class – the exciting 8- to 10-inch Dob market and the wealth of models the amateur can choose from.

CHAPTER FOUR

Getting Serious

Is there such a thing as the ideal 'scope? What do we mean by *ideal*? Well, perhaps it's an instrument that has enough light grasp to bag thousands of deep sky objects and enough resolution to tease out the finest details from lunar and planetary observations. Such a telescope ought to be portable enough to set up and cool down fast enough to make observations worthwhile on most clear nights. Many diehard observers reckon an 8-inch or 10-inch (20- or 35-cm) Dobsonian comes pretty darn close to that ideal for reasons that we will supply in this chapter.

A Truss Tube for the Masses

The Meade LightBridge range, introduced in 2005, can rightly be credited for bringing affordable truss tube Dobs to the masses. The series originally included instruments in 8-, 10-, 12-, and 16-inch apertures in "standard" and "deluxe" versions. After launch of the product range, the 8-inch instrument was discontinued, and all the Lightbridge 'scopes were sold in their deluxe form.

The quality optics for these 'scopes are manufactured by Guan Sheng Optical in Taiwan. Each one comes with a built-in 2-inch Crayford focuser, two cylindrical segments, housing the primary and secondary mirrors,

N. English, *Choosing and Using a Dobsonian Telescope*, Patrick Moore's Practical Astronomy Series 1, DOI 10.1007/978-1-4419-8786-0_4,

and a laminate-covered rocker box. The truss tubes are attached in pairs and are held in place by large thumbscrews. The whole instrument can be assembled without tools, although you'll still need to keep a flathead screwdriver handy to tweak collimation before use.

The primary mirror rests on a nine-point floatation mirror cell with support distributed by means of three groups of three pads across the rear of the mirror. Three padded restraint clips set 120 degrees apart at the front of the primary prevent it from falling forward or shifting position while moving it across the sky. The mirror cell permits the user to adjust the tilt of the primary by means of three screws accessible at the rear for tilt adjustment, while three other screws lock the tilt mechanisms in place.

To facilitate collimation, the factory provides a center reference mark on the primary mirror. Incidentally, this mark is located at the center of the mirror in an area obstructed from incoming light by the secondary mirror, and so this mark has no impact on the optical performance. The front aluminized surface of the primary mirror is protected by a stylish lightweight molded plastic slip on cover. The actual fit of the elliptical secondary mirror in both axes to be very close to nominal. This appears to be a very well thought out telescope.

All the Lightbridge Dobs are now provided with a 6 Volt DC-powered cooling fan positioned so that it draws air across the back of the mirror, thus helping to draw heat away from it. Some folk only use the fan initially to help acclimate the mirror to local temperatures. This should take about an hour of taking the telescope outdoors. The cooling fan set includes a corded battery holder that accommodates four "AA" batteries, and the cord plugs into a female socket. Including a fan with a telescope with this large a mirror is always a good thing, but that's not to say that there are no issues with it. Curiously, the fan has no protective grid, so it's important to keep loose objects away from the area of the carriage situated below the optical tube. What's more, the fan is made of fragile plastic, and so is susceptible to impact if the telescope is set aside upright.

The base comes flat-packed in several pieces – two side panels and a front, a base board and ground board, and a azimuth bearing, which is comprised of three different parts. The assembly took about 20 min and was very straightforward, with clearly written instructions. The sides and front are screwed on to the baseboard, and the azimuth bearing gets sandwiched between the ground board and base board. The base board and ground board are held together by a hex bolt with a large knob affixed so that it can be hand tightened – a nice touch. This knob controls how

Fig. 4.1. The Meade 10-inch LightBridge truss tube Dob (Image credit: Telescope House).

much friction the azimuth bearing gets. The next step is to place the bottom part of the optical tube (housing the primary mirror) in the rocker box and attaching the truss bars to the upper part of the optical tube assembly (the bit containing the secondary mirror) (Fig. 4.1).

The focuser is a well designed dual speed Crayford – a necessity on a 'scope with such a shallow depth of focus. The supplied eyepiece – a 2-inch Meade 26 mm QX – delivers a bright, fully illuminated field with a magnification of 49× and a field of view a shade under 1.5 degrees.

The 10-inch f/5 Lightbridge ($599) was put through its paces during an Easter vacation to a favorite dark sky (magnitude 7 skies) located about 10 miles north of the small town of Oban, on the west coast of Scotland. This delightful inlet blows onshore breezes each evening, which often abate after midnight to yield pristine northern skies that would take your breath away. The telescope took about 45 min to cool down to ambient temperature – with the primary mirror fan switched on – when taken out of doors. Collimation was way off, but this was fairly straightforward to remedy by aiming the 'scope at the second magnitude star, Polaris.

Repeated assembling and disassembling revealed that collimation holds up fairly well for casual deep sky viewing at low powers, but for high resolution work, it required frequent tweaking.

Inserting the supplied QX eyepiece, you can get very nice on-axis views, but as soon as the stars were placed about one third of the way out, the nice tight stellar images became puffed up and elongated from strong field curvature and coma. Fortunately, there was a set of quality ultra-wide angle eyepieces on hand that produced much better off-axis images. Star testing the 'scope revealed smooth, concentric rings both inside and outside focus. Just detectable was a very mild amount of astigmatism at high power. Movements in both axes were smooth enough to manually track objects fairly easily at magnifications up to 200×. Only on the rare occasions using still higher powers did some problems occur getting the right level of tension in the altitude and azimuth bearings to move the 'scope satisfactorily.

The 10-inches of aperture is enough to keep an observer happy for a lifetime. Observing the first quarter Moon, the Meade Lightbridge served up stunning amounts of detail. Indeed with Luna, its light-gathering power was sometimes overbearing. Indeed, you might be more comfortable using a neutral density filter while enjoying low power lunar vistas. Open clusters such as the Perseus double cluster and the trio of fainter clusters in Aurigae (M36, M37, and M38) were gorgeously rendered. Saturn was awesome, too. The Cassini division was etched into the rings like someone had used a black marker pen. Cloud details on the Saturnian globe were absolutely spellbinding during moments of good seeing, with hints of ovals, swirls, and subtle colorations glimpsed near the planet's poles. The details on view were, quite simply, in a different league to anything seen in any 5- or 6-inch commercial refractor. The only snag, of course, is that these periods of good seeing 'scope. Even with the fan turned on, thermal issues were quite often a problem, owing to the open, truss tube design. Throwing a light shroud around the instrument seem to offer some improvement in this regard and also cut down on stray light.

Later that year, the 'scope was set up to look at some of favorite globular clusters. M13 was spellbinding, as was neighboring M92. These fuzz balls that resolve poorly in smaller telescopes really open up with this size 'scope. Seeing a fully resolved bauble of stars floating across the field at 150× is an awe-inspiring sight and a faithful reminder of why aperture really does rule on these faint objects at least. The Meade 10-inch

Lightbridge is great value for money and an instrument that can be recommended to anyone wishing to buy an economical light bucket.

In 2008, Sky Watcher launched their own range of portable Dobsonians. Although the company had produced some fine closed tube Dobs for many years, these new Sky Watcher 'scopes had some delightful features, most notable of which is the flex tube design that can be thought of a truss tube design that needs no dismantling! The range includes instruments in the 8–12-inch range.

David Hartley, an experienced observer from Ramsbottom, England, had this assessment of the Sky Watcher Skyliner 200P Flextube Dob: "I am a visual planetary observer," he said, "and so I wanted a 'scope that would cool to ambient temperatures very quickly, have at least 8 inches of aperture with a small secondary obstruction, and be portable. The Sky Watcher 8-inch f/6 Dob seemed to fit the bill perfectly, so I ordered one, put in a cooling fan, flocked opposite the focuser, collimated, and began observing. I was instantly surprised by how quickly the 'scope cooled and by how good the optics were. The star test either side of focus was very good, and the 23 percent central obstruction was small enough to ensure sharp focus at very high magnifications. I have seen airy disks around stars at ×300 – that's a first for me in a 'scope of this size (Fig. 4.2).

"Amazingly I have collimated this 'scope only three times in fourteen months – but they were all down to me fiddling with the 'scope, otherwise normal movement does not affect collimation. The only issues I had was the center spot on the primary was not in the center, and some of the screws started to rust when I stored it in the garage. It has given me some superb views of Mars and Saturn over the 2009/2010 opposition (see attached drawings) and even a nice photo of Jupiter with a simple webcam. Overall it is superb value for the money and a very rewarding telescope" (Figs. 4.3 and 4.4).

How important was the addition of a cooling fan to the telescope's performance in the field? "The cooling fan made a huge difference in my opinion," David said. "I have always thought our UK skies were poor and I got less than satisfactory views until I installed the fan. Without the fan it takes ages for the mirror to truly cool, and even then the boundary layer and tube currents create instability ay high magnifications. The fan (a £5 computer fan) was a revelation in terms of cool down and producing a steady image – I observe with the fan on. It is 100 percent worth the investment."

Fig. 4.2. The 8-inch f/6 Sky Watcher FlexTube Dob (Image credit: Ken Hubal).

Fig. 4.3. Jupiter captured through the Sky Watcher 8-inch FlexTube Dob (Image credit: David Hartley).

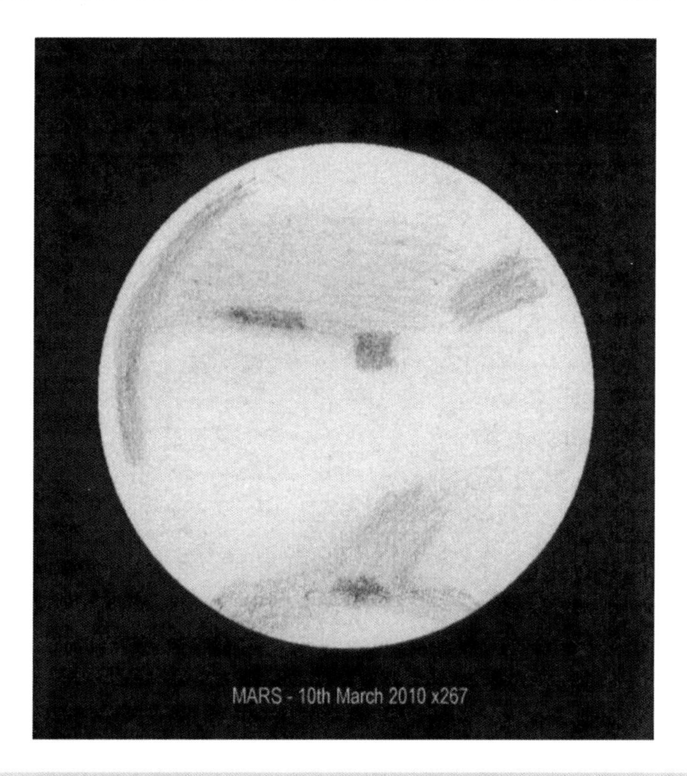

Fig. 4.4. Some images drawn and captured by David Hartley showing the planetary prowess of the Sky Watcher 8-inch f/6 FlexTube Dob (Image credits: David Hartley).

Fig. 4.4. Continued

A Super Cool Dob

Mag 1 Instruments, based in Chilton, Wisconson, offers an altogether different vision for the Dobsonian reflector. Company founder Peter Smitka abandoned the typical cylindrical optical tube and Lazy Susan cradle mount and replaced it with a telescope that sits inside a hollow sphere that moves rather like a ball-and-socket joint. His marvelous construction is called the Portaball. Mag 1 offer three different-sized Portaballs in 8- and 12.5-inch apertures. By far the most popular is the 8-inch f/5.5 instrument ($1,380 plus shipping), which has put a smile on the faces of many happy customers. To see why, read on (Fig. 4.5).

One enthusiastic owner said, "I have had a number of 8-inch Dobs over the years, but nothing that matched the quality of the image served up by the 8-inch f/5.5 Portaball. I have the version with 2-inch focuser and curved spider. Though I've only owned the instrument for less than a year, I have seen things I never saw in my previous 'scopes. The views in many ways remind me of those served up by my Apo refractor. The 'scope delivers great contrast, courtesy of the Zambuto primary mirror and the small (20 percent obstruction) secondary. It is a fantastic all-around performer and shows substantially more detail on the Moon, Jupiter, Mars, and Saturn

Fig. 4.5. The 8-inch Mag 1 Portaball (Image credit: Mag 1 Instruments).

than many premium 5- or 6-inch refractors I've compared it with. For example, the Portaball can resolve the Encke minimum in the A ring on nights of good seeing. The larger craterlets in Plato on the Moon appear as craters, not just bright "patches" with my smaller 'scopes. The ball mount is a real joy to use, too. It is quite literally simplicity itself! I would recommend getting a tracking platform for this 'scope if planetary observing is your thing. That said, manual tracking at 300× or 400× is certainly doable, but I tend to waste half my viewing session bumping and waiting for the image to settle at these powers. I would also recommend purchasing a light shroud for the telescope, as I have found it cut down on stray light as well as ground thermals on warm summer evenings."

The Mag 1 8-inch Portaball sure is a charming instrument that exudes quality from tip to toe.

A Dob That Finds Things

What about a big Dob that won't break the bank and allows you to find any one of 14,000 objects, the locations of which are stored in a small book-sized computer? Javier Medrano from Minneapolis has thoroughly evaluated the Orion XT 10i over a period of two years and was kind enough to offer his opinion on its optical and mechanical performance. "After owning a few small reflector telescopes and giant 20 × 80 binoculars," he says, I felt the need for a bigger aperture telescope, but still manageable in size, and after some research, I settled on the Orion Skyquest XTi 10 inch Dob with Intelliscope Object Locator I. The telescope came in two well packed boxes, and everything was in good shape. Unpacked, the telescope is 29 lbs and the Dobsonian base, assembled, weighs in around 26 lbs. It's easy to move the telescope in two parts.

The optical tube has two altitude bearings like "ears" that rest on the bearing cylinders of the base and makes it both easy and secure to hold the optical tube while walking with it. I did have to be careful when lifting the optical tube from the base, particularly after a night where dew makes the tube slippery. Because of the one-piece nature of the telescope tube, it is, without doubt, bulky. That said, I tossed it regularly into the car, together with a heavy-duty Orion soft bag, and it holds collimation well even after long trips. The 'scope comes with an impressive package of accessories; two Orion Plossl eyepieces – 25 and 10 mm – of decent quality, a top of the line 9 × 50 right angled finder 'scope nicely color matched with the optical tube, a 2-inch single-speed Crayford focuser with a 1.25-inch adapter, two encoders for the base, the Intelliscope handheld locator (I got the new version that shuts off at 50 min if left untouched and not the annoying 15 min on the older version), cables, instructions, and hardware.

"I am not what you'd describe as a "handyman," but within an hour and a half, everything was set up.

The focuser is a fine, single-speed unit, but I have no complaints with it whatsoever; it has big knobs with rubber channels, and I can focus perfectly even with thick gloves. The focuser is collimatable, too, and also has a focus-locking knob. Movements are smooth without any "jerkiness" at all, even on cold nights (Figs. 4.6–4.8)

The telescope comes with a hard plastic dust cover, with no aperture mask on it. It fits fine in mild and hot weather, but it is a little loose on cold nights. A tray attaches to the base where you can place three 1.25-inch eyepieces and one 2-inch eyepiece. There is also a handle to hold the base

Fig. 4.6. The Orion XT 10i fully assembled and ready for use (Image credit: Javier Medrano).

while transporting it. The cutouts of the sidewalls are a very good idea, as they save weight and I use them as the two "handles" to move the whole assembly together very short distances. Careful, it is heavy in one piece; I can carry it, but why risk bumping into something or worse? I find it is faster to move it in two pieces (base and optical tube assembly), especially with any stairs involved.

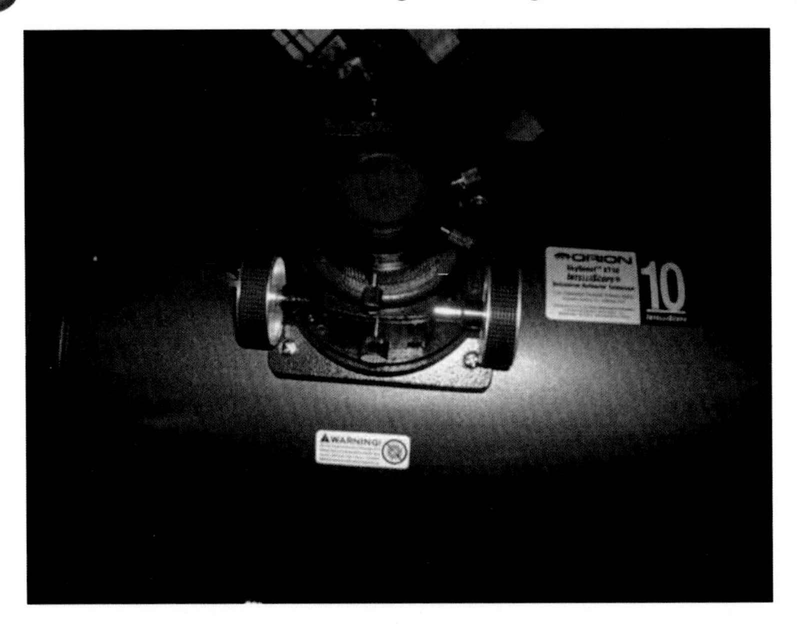

Fig. 4.7. The basic but functional Crayford focuser on the Orion XT 10 (Image credit: Javier Medrano).

A cooling fan for the primary the mirror is not provided; I got the fan on mine later on. It takes no time in the summer to cool down, but up to one hour in cold weather. You cut that time in half with the fan. The mirror cell comes with pre-drilled holes for aftermarket fans. One thing that bothered me is that I had to buy the handle controller mount as an extra item. It is not essential and not very expensive ($20), but adhesive Velcro is all the original boxes had to affix it to the base. The position of the Intelliscope mount is very convenient while in use. It is high up and slightly tilted, and I just leave it there even when the 'scope is stored away indoors (Fig. 4.9).

I use a drummer style round seat, and I can be at the eyepiece, comfortably seated in every position I can point at, from very low horizon targets to the zenith. I believe the size of the 'scope is just right for viewing in comfort for extended periods of time. I am 5′ 8 tall, so any standard chair will work just fine. I find you get the most pleasure observing while seated. It's completely relaxing (Fig. 4.10).

The 9 × 50 finder was easily adjusted via two nylon thumbscrews and a spring pivot support that keeps it snugly in place. It is of very nice quality, and I use it on my other refractor telescopes. It can even be focused.

Fig. 4.8. A close up of the encoders that enable "PushTo" capability on the Orion XT10 (Image credit: Javier Medrano).

The views on mine are crisp and clean, with long eye relief. You have 5 degrees of view in your cross hairs. I do not have a Telrad, but I don't think you'd really need one either. Besides, I prefer to see through the finder to see enough stars to know where I am.

Under dark skies it works like a dream. I do not have problems seeing the crosshairs in the dark. Collimation came next. Using a laser collimator,

Fig. 4.9. The nicely finished interior of the optical tube (Image credit: Javier Medrano).

Fig. 4.10. Ready for first light. The Orion XT10i awaits star light (Image credit: Javier Medrano).

the secondary was just a hair off; it can be adjusted with three setscrews at the spider (now I have Bob's Knobs, a must of easy adjustment); the primary mirror is factory center-marked, and the primary needed more adjustment, as expected. The mirror cell has three wide collimation screws and three smaller locking screws next to them. Needless to say, the collimation process is easy. In addition, no tools are necessary for the primary. I find that there is no need to collimate the telescope at all if it has not been moved. I have checked the collimation many times during observing sessions and it is always good once you perform the initial collimation procedure after setting up.

The first object I used to test the telescope was a familiar one: I pointed it at the double cluster in Perseus without resorting to using the Intelliscope. I used the supplied 25-mm eyepiece included with the 'scope (×48 magnification). Movements were very smooth in both axles, not a trace of 'sticktion' whatsoever, and you can control the tension for the altitude, turning the right side clutch knob. The 'scope is a little bottom heavy, so I usually resort to tightening the right side clutch when tracking at high magnification or viewing the zenith. The left side knob serves as a security lock for the OTA-base, and also as the OTA-altitude encoder lock. It must be fully locked for the altitude encoder to work. The knob located in the underside of the aperture end makes moving the telescope much more 'natural' than pushing the OTA with your hands at normal angles. As also expected, the movements in azimuth of the base get more difficult if you are pointing to the zenith. You need two hands then for fine pointing, but is still smooth and much better than my old Orion 6 inch Dob (Fig. 4.11).

Regarding optics, the double cluster was incredible. A completely new world just opens up in front of your eyes! There was some coma on the outermost portion of my field of view, but nothing dramatic at all, and nothing that a coma corrector couldn't effectively remedy. I am very, very happy with the optics in this unit. I am actually very fond of the supplied 25-mm eyepiece. It gives me lovely wide views of open clusters, and it's a nice magnification for scanning with this telescope. In general, I like 20- to 32-mm eyepieces for this 'scope. For observing smaller and fainter globular clusters and planetary nebulae, I use eyepieces in the 12- to 17-mm focal length range.

I have looked at the Moon and Jupiter with the Orion XT10i, and because this telescope collects so much light, I find it useful to employ a Celestron Moon filter. With a 2× Barlow and the supplied 10 mm Plossl it yields a magnification of 240×. Observing the first quarter Moon was a magical experience. Large craters and tiny craterlets inside of more craters…crisp details everywhere, ridges, valleys, dunes, and flats. On the best nights I can crank up the power to 375× with this 'scope for lunar observing without hesitation. Indeed, I watched the LCROSS Moon

Fig. 4.11. The handy holing knobs make moving the Orion Xt10i a breeze (Image credit: Javier Medrano).

impact on June 19, 2009, using these high powers, although tracking is tricky at these magnifications.

Next was Jupiter. I could see four bands at ×240, not with too many details on festoons, though. I did not have very good conditions, but I spent a good hour on this target. I expected slightly better views of Jupiter....Only once or twice after that, I have been happy with the Jupiter views and could see nice details at 300× to 350× when conditions improved. Most of the time, I can't go over 250× without image degradation. Focusing is very tricky at high power with this 'scope (a consequence of its hollow depth of focus). A blue filter does seem to help bring out more a surface detail, but I need darker skies and better seeing to really get the most out of this 'scope.

Regarding the Intelliscope locator, if you are familiar with the instructions and can locate and name the bright stars in the sky, it is a simple operation. First turn it on. (Pressing the ON button repeatedly rotates the levels of the intensity of the screen and buttons backlight). Then it asks you to point the telescope vertical. Do it and press ENTER. Then choose

Fig. 4.12. A close up of the brains of the Orion XT10i Intelliscope (Image credit: Javier Medrano).

the first alignment star from the option list that appears on the screen. Next you highlight the star you want with the arrow buttons, center it on the eyepiece, and press ENTER. You then repeat the procedure for a second star. If your hardware has been correctly installed, and you have correctly found and centered the two stars at the eyepiece, your "warp factor" (the number that the handheld shows you after you complete each two-star alignment) should be less than +/− 0.4. Repeat the alignment if this turns out to be higher (or not; .06 will get you close) (Fig. 4.12).

When you enter "M31" in the object locator, for example, two arrows and two numbers appear. Just move the telescope until both numbers are zero, and look trough the eyepiece to see your target. With a bit of practice, you can reduce the warp factor to about +/− 0.2 almost every time (with a lot of zero warps!!), which means that 99 percent of the objects I dial in at the handheld are right in the middle or very close with the 25-mm eyepiece field of view; that is just great – it works all the time. It even works quite often with higher power eyepieces (12–15 mm) that I occasionally use. That's quite accurate!

I have noted that on cold nights, the text lines on the object locator are barely readable at all. On some of the nights this winter, here in Minneapolis, the temperatures fell to −31 degrees F, and under those extreme conditions, the numbers also disappear. When things get that cold, I can only venture out for one hour maximum. The 9-volt battery lasts 25–30 h, I would say, at 2-3-4 h per session, at mid to low brightness.

I have had only one glitch with this technology. At one point, the alignment factor was giving me some trouble, so I called Orion. Right away, they told me to check the altazimuth encoder assembly. I did: the center screw of the base was slightly loose, so the magnetic encoder disc was not moving with the rotating base. After I fixed that I got back my usual +/− 0.2 Warp factors.

The person I talked to did mention that alignments are done better if you avoid Polaris as a second star. In addition, they informed me that I should try to align with stars on the same side of the meridian, seems to work better, people have reported. In my experience some star combinations give better warp factors than others. You have to play a little with some combinations until you find which ones are good for that time of the year – a few tries will do it. To get the best out of the object locator, it pays to be as accurate as possible in your star centering activities; this is the key. Now, it is always on target. I was having a hard time star hopping from my urban skies with only 30–35 stars seen with naked eye on a good day. There are 12 "tours" available on the locator, one per month; they show some 20 or 30 targets, more than enough for a couple of viewing sessions, but those targets are fixed and the tours do not take the local time into consideration, so some of them are below the horizon or not yet up when you go look for them. It also can find the planets for you, but you have to input the date by hand. I asked Orion and there are no updates available for the software or tours.

As well as storing the coordinates of all objects on the three catalogues it has (Messier, NGC, IC, plus around 800 double stars), as well as brief descriptions on important targets, the object locator also enables you to enter additional celestial coordinates not normally stored in the system.

You can store 99 of your own objects by RA/DEC location. One of the best features of the remote is the ID function; it identifies what you are looking at. Remarkably, it also does this fairly accurately. I love to scan at low powers and it's nice to know right away, what it is you have stumbled across. I usually have *Stellarium, Starry Nights,* or *The Sky* software in my laptop to find objects to dial into the Intelliscope controller. I also got the special Orion cable to connect the locator directly to the computer, and used in this way it shows where are you are pointing the telescope at on the screen map. This can be done also via Bluetooth instead of cables with a Bluestar adaptor, and I have read of some people doing it with a pocket PC and the *Sky Pocket* edition.

There is an option for improving the accuracy of the alignment: if your object is a little off: center it on the eyepiece, press the "Function" key, then "Enter," and a new warp factor is given for the session. Usually I do not need this feature, but it is there in case I am too lazy to re-align from scratch. I had some issues with this feature, and some new warp factors were worse than the originals, forcing me to realign from scratch (it loses some accuracy after 3 or 4 h, probably because I move slightly the base when manually slewing to find target. Other nice targets that I have enjoyed a lot in this 'scope are: the Orion Nebula with the trapezium well resolved at very low power. At high power I have split the doubles E and F easily; green tint all over the nebula, gas and dust lines are very evident. M31 is "brutal," showing M32 and M110 very well. The Ring Nebula is spectacular in this thing, perfect circle, super bright, just a joy. M35, M36, 37, and 38 are some of my favorite clusters, wide field views in this telescope are the best. M45 is a light show with hints of nebulae, M44 another joy....All of the clusters at Cassiopeia and Cygnus are great for scanning. Globular clusters like M13, M92, M3, M22, M19...are bright and very well defined; none of my other telescopes came even close at all. I must admit, though, that I went from viewing fuzzy blobs to cleanly resolved star clusters (my previous 'scopes being 4.5-inch and 6-inch Newtonians).

A tour of Sagittarius with my Orion XT10 is a visual feast. The colors in Albireo are gorgeous. After almost a year now, I am very pleased with this telescope. In my opinion, this is great for deep sky objects such as nebulae, open clusters, and globulars. Planetary nebulae such as M57 and the Ghost of Jupiter are fantastic, too. I can see galaxies galore from dark sites. I toured the Virgo Cluster in April 2010 and was frankly amazed at how many galaxies I could see in dark sky conditions, even pushing the magnification. At low powers, I could frame some times 3-4-5 galaxies in the same field of view. M51 is delightful, and the spiral structure is seen clearly with direct vision. M81-82 and the Leo Triplets are a fantastic sight.

Though this telescope is very decent on most double/multiple stars also, I don't think it is great on planets, but I must say that I prefer to see planets through this telescope more than through any of my achromatic refractors (the biggest I own is a nice specimen Orion SVP 120 F8). That said, I have seen Saturn and Mars on 2010 during their oppositions from dark skies, and I could clearly see the polar caps on Mars and a very pleasant view of Saturn (the rings are almost perpendicular so I cannot say too much about ring divisions…). The scale is "smallish" on the eyepiece due to the short focal length of the 'scope, especially in Mars, even at 300X power. That is also due to the far opposition of Mars this year. Venus is featureless, but the phase can be seen very well, but in all honesty you almost need a filter to cut the light down!

Regarding wide fields of view and scanning the star-rich constellations, I also prefer this 'scope to my own Orion ST120 F5. The gains on contrast on the smaller refractors simply cannot compare to the light-gathering power of the 10-inch aperture. My apologies to the refractor lovers, but at least for me, I prefer to use the XT10i on all kinds of objects.

One night I compared the view of my telescope with a 12-inch Orion XT Classic, owned by a member of the Minnesota Astronomical Society. I must say that views of the Orion Nebula through the 12-inch were marginally better than my smaller Dob from our dark sky site, but I'm not sure if the increase of size and weight is worth it for everybody. My 10-inch could see much the same detail. Yes, the 12-inch delivers bigger and brighter views, but again, only slightly better. The difference in weight is more than 30 pounds, however, and both the optical tube and base are significantly bigger. If you are not lazy and can handle the extra weight, then, by all means, go with the 12-inch. But I reckon the 10-inch Dob is the perfect compromise. I must say that I would buy it again. It is a nice 'scope. It calls me every clear night, because the views are good, the set up is fast, alignments are easy (or you can have a star-hopping session without electronics), you can be seated comfortably when viewing at any angle, and it almost always delivers the "wow" factor we amateurs crave.

Javier has described this popular econo-Dob in great detail, and it is broadly consistent with a few other models in this genre. The XT10i ($700 plus shipping) gives you decent optics, a stable mount, and all the pleasures of PushTo technology for a very respectable price. Similar models marketed by Zhumell and GSO can be retrofitted with the same PushTo technology. In the hands of an amateur with the kind of positive attitude to observing as Javier's, you can't go wrong with one of these light buckets. In the next chapter, we'll go in a totally new and ambitious direction as we make the case for the "planetary" Dobsonian.

CHAPTER FIVE

The Planetary Dobs

A revolution, by necessity, must permeate all rungs of society and evince changes to that society that are palpable and long lasting. The Dobsonian "movement" didn't become a "revolution" until it could reach out from its traditional hinterland of deep sky nirvana and begin to make inroads into niches traditionally dominated by other kinds of 'scopes. We're thinking, of course, of planetary observing, where the smaller aperture, high-quality refractor has traditionally dominated. To compete with the refractor (or Maksutov Newtonian for that matter) a reflector ought to have a high focal ratio to allow comfortable high power views to be achieved. It ought to have exceptional contrast by minimizing stray light as well as possessing a central obstruction less than 20 percent. Such a telescope, optical theory suggests, ought to behave as a superb lunar, planetary, and double star 'scope, rivaling a refractor costing many times more.

Right off the bat, that's a tall order for a commercially produced Dob, especially since it is well known just how fine a good refractor serves up images of these objects under the right conditions. Yet, there are now 'scopes on the market that claim to give refractor or Mak users a run for their money. This chapter takes a closer look at some these instruments – invariably 6–8-inch Dobs – and explores whether they can deliver on their promise of rivaling the views through refractors. Now, it's true that a larger instrument, even a modest 8-inch f/6 Dob, would wipe the floor

N. English, *Choosing and Using a Dobsonian Telescope*, Patrick Moore's Practical Astronomy Series 1, DOI 10.1007/978-1-4419-8786-0_5,

of any "super" 6-incher! But that's not what these Dobs are about. Even a 6-inch f/10 or f/12 'scope is still lightweight and fairly portable. The small mirror guarantees a relatively quick cool down time, and if you install a fan, you can be up and running in 30 min after taking it out of a warm room indoors.

Such a telescope is quite appealing to a lunar and planetary enthusiast who wants good results. It pays also to remind ourselves of the many attributes of high focal ratio 'scopes. First off, their generous depth of focus makes precise focusing easier and thus helps maintain a more stable image. High F ratio 'scopes have a large diffraction limited field of view, so that even mediocre eyepieces work extremely well with them. From a practical point of view, they are easier to collimate and more forgiving of errors in alignment compared with faster 'scopes. With these facts in mind, let's take a closer look at some prime planetary Dob candidates.

As was discussed in Chap. 2, when the central obstruction of a conventional reflector approaches about 20 percent of the linear diameter of the mirror, the images become much more like those delivered by the refractor. The effects of the diffraction spikes are far more subdued, and the contrast becomes noticeably better. First in the lineup is Star Gazer Steve's 6-inch Deluxe Reflector kit. For $379 plus shipping you get a chance to learn about and build a powerful telescope that you can use the same night. The kit includes an easy-to-follow, step-by-step video and mounting parts of high-quality birch plywood, centered around the optics of a quality 6-inch f/8 parabolic mirror (and he can produce higher F-ratio if so desired.) Owner reports are typically glowing!

Vic Graham from Wichita, Kansas, described his experience of building one of these from scratch. "I decided to buy a telescope from "Stargazer" Steve Dodson in early April. When I asked how long the wait was, Steve told me it would be 2½ months. I sent my payment expecting delivery in late June or early July. The kit arrived 4½ months later at the end of August. There were times I was very frustrated with the delays, but Steve remained polite. It's obvious from our communications Steve has another job, and building telescopes is what he does on the side. I suppose a wait can be expected when purchasing a telescope that isn't mass-produced and sitting on a shelf somewhere. I've read of similar experiences when buying a Discovery or Obsession telescope. The package arrived in a single long box that looked like it had endured some abuse. The contents were packed well, though, and nothing was damaged. The main and primary mirrors were safely stowed in their rightful places in the tube. The kit includes everything needed to build the telescope except for a few tools. At a minimum the builder needs to have masking tape, a hammer,

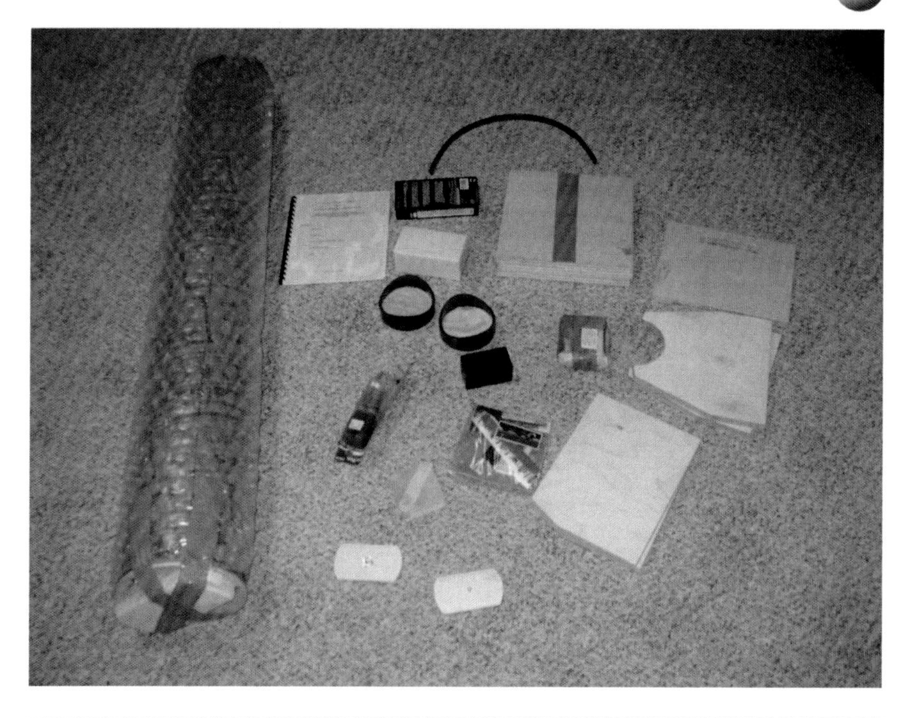

Fig. 5.1. Star Gazer Steve's planetary Dob (Image credit: Vic Graham).

a Phillips screwdriver, a standard screwdriver, a ruler or tape measure, and an Exacto knife or similar suitable for cutting thick cardboard. I also used a file and a drill. Construction didn't require a high degree of skill, but I had to improvise to overcome a couple of problems (Fig. 5.1).

The tube is made of a 7.5-inch diameter, 52-inch long concrete form. The base is made of ¾ inch birch plywood. A thin piece of aluminum is bonded to the bottom of the rocker box and glides on three Teflon pads mounted to the ground board. The kit includes a very helpful instructional video (in VHS format) that runs one hour and should be viewed at least once before starting construction. You might need to consult the video a couple of times for more difficult steps. Seeing the steps carried out proved to be a tremendous help, especially when installing the secondary mirror. The instruction manual suggested finishing the telescope after building it, but if you don't want to work around the mirrors and focuser or disassemble anything, paint the tube before you start work. You can also stain the rocker box and ground board before attaching the Teflon pads and completely assembling the base (Figs. 5.2 and 5.3).

Fig. 5.2. Getting to grips with the base (Image credit: Vic Graham).

Fig. 5.3. Spray painting the OTA (Image credit: Vic Graham).

Construction requires cutting a hole for the focuser and a door at the bottom of the tube. The door doubles as access to the tube balance weights and a vent for thermally stabilizing the mirror. Cutting is made easier by the perforations drilled into the tube. Holes are pre-drilled in the tube for attaching the bearings, focuser, and Rigel finder. It was little trouble installing the focuser. Two of the holes pre-drilled for the focuser didn't match the focuser. If the holes are just a fraction off, use a drill to widen them, then fasten the focuser. One of the biggest challenges was installing secondary mirror the right way, so watch that portion of the instruction video a couple of more times. The process involves using masking tape to make an "X" across the bottom of the tube and using the "X" to align and adjust the secondary while looking through the focuser. The spider should fit very tightly in the tube, and it is hard to make fine adjustments. Be very careful and take the time to get it correct. After the spider is in place, it's permanently fixed with some fast-drying epoxy glue (Figs. 5.4 and 5.5).

Once the secondary is finished the mirror cell is reinstalled and collimated. The manual provides very good instructions and diagrams to help. You can use a film canister with a hole drilled in it to complete collimation. The mirror cell doubles as a ballast box for balancing the tube.

Fig. 5.4. The weights go inside here (Image credit: Vic Graham).

Fig. 5.5. A hole for where the eyepiece goes (Image credit: Vic Graham).

A couple of large, heavy, threaded studs and some enormous nuts are already installed in the kit. A couple of more studs and several extra nuts are included for balancing heavy accessories. You would do well to install one more studs and a couple of nuts to balance any eyepieces you bought for the telescope (Fig. 5.6).

The rocker box is simply nailed together and attached to the ground board by a screw with a rotating collar. The finished telescope weighed in at 29 pounds. The center of gravity is low, making it stable even on a relatively small base. The motion is very smooth and accurate. The project takes 8 or 9 h including paint and stain. The Stargazer Steve 6-inch telescope is a good value. It's a good product for people who want a unique telescope and are willing to spend some time building it. It can be carried outside in one piece and easily moved around a yard, and has a very useful Rigel finder. The balancing system is innovative. Its motion is very smooth and easy and is easy to aim and track. It gets many compliments from experienced observers. They seem impressed

Fig. 5.6. The completed tube base together with balancing weights (Image credit: Vic Graham).

with the quality of the views. One night a guy was unsuccessfully trying to split the double double with a home-made 14 inch Dob. This author was able to show him all four stars. He was impressed. I had another guy try out his new 9 mm planetary eyepiece on Jupiter. He was very impressed and spent some time looking at Jupiter through the 'scope." That's not surprising given the small (20 percent) central obstruction due to the secondary mirror.

"My dislikes are few," he said. "Since I bought my 100ED refractor, I would prefer a smoother 2-inch focuser. Over time the base has developed a slight wobble on solid surfaces I haven't figured out yet. I haven't spent much time on it, and I've managed to solve the problem by observing in the grass or using a thin, firm carpet sample. I didn't like waiting so long for my 'scope kit to arrive. If you still want me write something up, just give me some more guidance. I believe a 6-inch aperture is very underrated, and I've enjoyed many hours looking through my 'scope and bagging dozens of deep-sky targets" (Fig. 5.7).

Fig. 5.7. The finished product (Image credit: Patrick Dodson).

A British Planetary Dob

Orion Optics, based in Cheshire, UK, has its own answer to the Apo killer. Called the OD 150 L, it's a 6-inch f/11 reflector with 25-mm (17 percent) central obstruction. Available in standard (1/4 wave specification primary) form for £399 (tube only) or as a deluxe model (with 1/6–1/10 wave figure) for £450–£500, these Dobs are easy to set up and even easier to use. This telescope, weighing about 33 pounds (15 k), was created to meet the demands of the discerning lunar and planetary observer, where set up time is a premium. If you've not seen an Orion Dob mount in person they are very solidly built, fashioned as they are from 10-mm thick alloy plate and eminently capable of securing the long 6-inch optical tube assembly. And there's no chipboard in sight!

Pat Conlon from Northern Ireland was kind enough to provide some answers to questions about the Dob, beginning with the moment he inspected the newly arrived package from Orion. "On opening the box and removing the 'scope I saw there was something missing," he said. "It had no finder, but a quick email to Orion sorted that out. When I was ordering the 'scope I had requested a thicker tube mirror upgrade to 1/10 PV and 2 speed focuser. This is a long telescope 154 cm and only 17 cm in diameter; it is optimized for high power viewing, although when you rack the focuser in, the body of the focuser intrudes into the light path (as illustrated in photos). I contacted Orion about this and they offered to mill a bit off it if I sent it back to them. I haven't taken them up on their offer yet. Out of the box, collimation was pretty good. I'm no expert but I couldn't find any real faults with the mirror and with a Strehl ratio of 0.994 this is to be expected.

On first light I was amazed how bright the images were for a 150-mm 'scope. This is because of the Hi-Lux coatings on the mirror, which are also supposed to last longer than normal coatings. Being mostly interested in double stars, I tried it out on a few of my favorites. First Iota Cassiopeia, a lovely triple star system, which the Orion had no trouble splitting with a 10-mm eyepiece giving 165X. It even showed a nice bit of color in the companions. Next stop was Castor in Gemini, which showed the stellar duo with a black strip separating them. From there, I panned the 'scope onto Theta Aurigae, a tough spilt because of the 4 or 5 magnitude difference between the components. I had to crank up the power on the Orion to 275× in order to yield a satisfactory split. It's a great performer on a variety of deep sky objects, too. The small central obstruction probably has a role to play here (Fig. 5.8).

I also had a chance to look at the Moon and planets. Lunar detail in this 'scope are simply wonderful. You can see craters inside craters as well as the shadows cast by the Sun inside them. Saturn was a sight for sore eyes, its rings thin but razor sharp. I could easily see a few of its satellites off to one side of the planet. Indeed, I even captured a hint of color in Titan's smoggy atmosphere at a power 275×. This high F ratio Dob takes magnification very well. Only when I Barlow my 6-mm eyepiece to give 550× do the images begin to get soft. This is a very good 'scope for the Moon and planets and would be even better if the focuser was a low profile type. I find it strange Orion did not do this, as it is advertised as a planetary 'scope. A few months ago, I had a chance to compare the

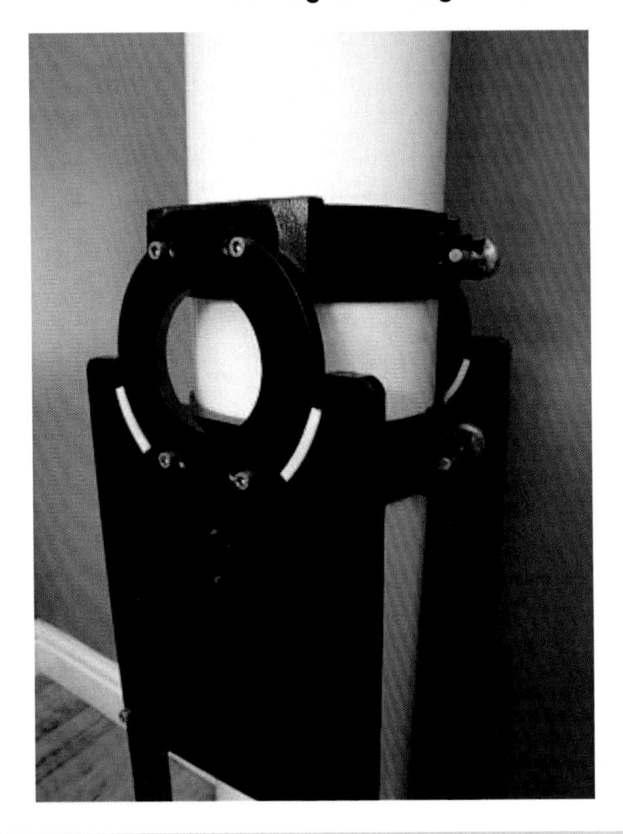

Fig. 5.8. The quality Orion Planetary Dob mount (Image credit: Shane).

Orion 6-inch f/11 to a Celestron Nexstar 6, and the former gave noticeably sharper and brighter views, although it was a night of poor seeing and cut short by clouds. I believe the Orion was more than a match for the Celestron. So, all in all, putting the issue with the focuser aside, Orion has produced an excellent telescope that is well worth the money in my opinion" (Figs. 5.9 and 5.10).

Some people think the Orion 6-inch f/11 Dob is expensive. But you have to remember that these are hand-built 'scopes that are made to order. The 1/10PV mirrors are nearly £300 and the Crayford focuser would cost over £200 on its own. Even when the extra cost of these luxury items is factored in, the price for the deluxe version with 1/10PV optics is still very reasonable. Shane Ward from Stockport responded to whether he thought the Orion Planetary Dob was worth the money.

Fig. 5.9. The high-quality Crayford focuser on the Orion planetary Dob (Image credit: Pat Conlon).

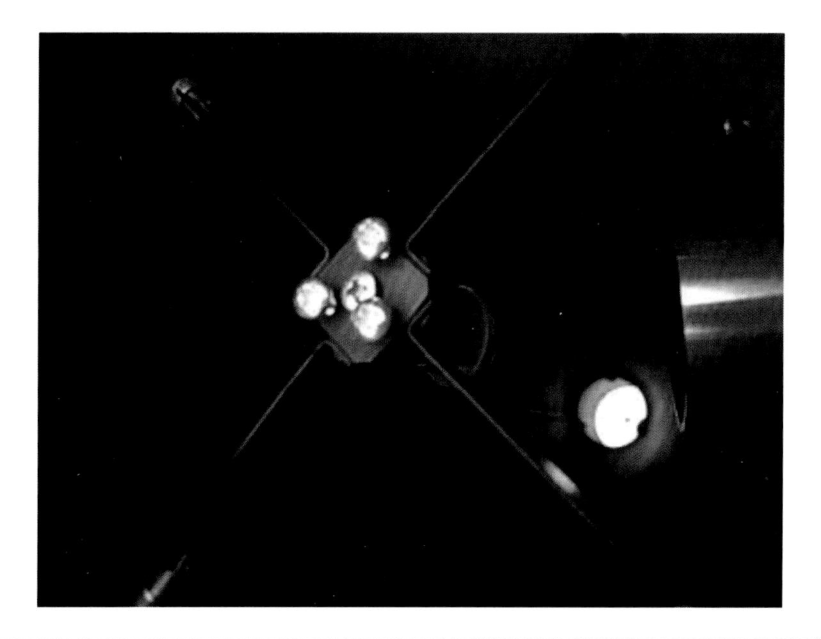

Fig. 5.10. Note the small (15 percent) central obstruction on the Orion Optics 6-inch f/11 Superplanetary Dob (Image credit: Pat Conlon).

"In answer to your question," he said, "I personally am of the view that the f/11 was more than worth the money, even though I bought it used. I think a new one with high grade optics would be a little expensive, but I got 1/6 wave P-V and HILUX for that price, so I am delighted. The contrast and detail on Jupiter has been superb recently and is far superior to the 120-mm refractor I had, and it even gives the 12-inch Dob a run for its money on most nights. It really is stunning to use. I love the base, but balancing it is a pain, and the friction brake is absolutely essential.

"I'd also recommend removing the Teflon blocks and adding felt or cork to the altitude bearings to give a little more friction. I think if, as a Dob owner, you accept you'll always be fiddling with it to make it better, then you'll be delighted, as this 'scope offers lots of opportunities for upgrading, including flocking, adding a fan, installation of a better focuser, and employing tool-free collimation knobs. I think some may be put off by the relatively high price tag of this 'scope, especially if you're ordering up a premium mirror. In truth, this 'scope performed well out of the box, but for me, owning a Dob is all about customization and improvement to your own tastes."

The Off-Axis Brigade

What if you could reduce the size of the secondary obstruction to zero? Yes, that's right, zero! Then you'd have images that rival a refractor in terms of sharpness and contrast. So-called off-axis (OA) Newtonians have come and gone over the years, but the current line, by all accounts, are super performers. Doug Reilly from the Finger Lakes area in upstate New York has gained extensive experience with a number of off-axis Newtonians. "DGM produced a line of successful off-axis Newtonians but are now no longer in production, though they do pop up on the used market," he said. "Dodgen Optical (Flagstaff Arizona), the shop that produced the optics for most of Dan McShane's OA telescopes has now introduced their own line. The Dodgen 'scopes are f/7.5, while the DGM OA 'scopes were always f/10. I've owned a 5.5 f/10 DGM and currently own a Dodgen Optical 6-inch f/7.5 primary mirror mounted in a tube assembly of my own design. Off-axis Newtonians are optically similar to traditional obstructed Newtonians that have been 'stopped' down with an aperture mask to remove the obstruction (and light scattering) of the secondary mirror and spider. OA mirrors are cored from large, fast, parabolic primary mirrors, each one yielding four off-axis mirrors. These 'scopes are quirky, but they do offer fantastic views. Collimation takes a little while to relearn; the principles are just a little different than a traditional Newtonian. Once you've got the knack of it, however, it's actually quite simple and requires nothing beyond a sight tube or a laser for the initial secondary alignment. Simplicity is nice! (Fig. 5.11)

Fig. 5.11. The DGM OA 5.5 inch on a simple Dobsonian-style mount (Image credit: Doug Reilly).

The optical boon of these 'scopes is contrast (which comes from the lack of a central obstruction) and sharpness (which comes from the extremely smooth figure of the high-quality parent mirror). They are truly fantastic in this regard, serving up a dark background sky, very little light scatter, and razor-sharp definition. I've compared the 6-inch to a 12.5-inch Zambuto-mirrored Compact Precision Telescope (CPT, a traditional Newt in terms of optical configuration) recently on M42, and the view was more pleasing in the off-axis 'scope even though its mirror is half the diameter! Indeed, three of my observational "bests" were in OA 'scopes. The OA5.5 yielded the finest image of Saturn I've ever seen, and the 6-inch Dodgen mirror gave me the best view of Jupiter in a (short) lifetime of observing that planet.

More recently, I was elated to actually see geological detail on the Martian surface in the 6-inch Dodgen. I've been looking at the Red Planet for years through every 'scope I have owned and have never seen anything behind the ice cap. I was blown away. The downside optically is a bit of coma that is worse on one side of the field of view. I don't find it objectionable if the optics are well-aligned. The longer DGM 'scopes are easier to collimate than the shorter f7.5 units produced by Dodgen, though the principles are the same. The Dodgen 'scopes are shorter and a bit easier to maneuver. Many DGM 'scopes came mounted on a very nice wooden Dobsonian mount with a high center of gravity. Are OA 'scopes worth the money? To me, the sharpness and contrast of off-axis Newts (which is world-class) far outweigh the learning curve for collimating the optics or the small amount of coma. And remember, though OA 'scopes are not cheap, they are still a small fraction of their optical peers, the apochromatic refractors" (Fig. 5.12).

Doug Reilly's sentiments seem justified especially in light of the significant difference in price between these larger off-axis 'scopes and the equivalent-sized apochromatic refractor. But what about the smaller instruments? Orion Telescope and Binocular brought out their own version – a 3.6-inch f/10 off-axis Newtonian – a few years back. But it

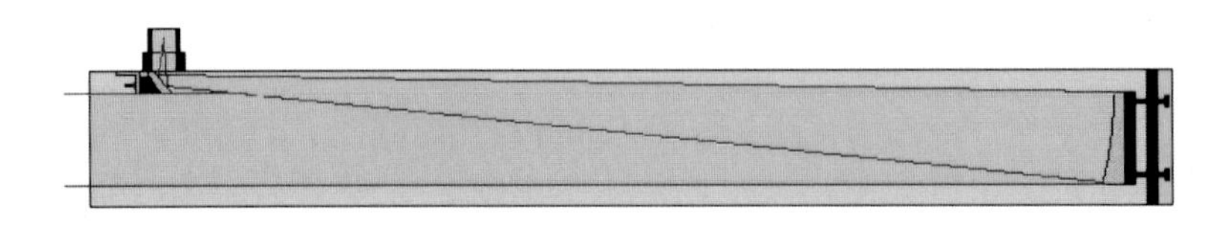

Fig. 5.12. Optical design of the DGM OA Newtonians. Note its totally unobstructed optics (Image credit: DGM Optics).

received mixed receptions. One reviewer quipped that although its optics were the equivalent of any 3.6-inch apo, it was easily out performed by a conventional 6-inch reflector. True, but surely the point of such an instrument is to produce images that are the closest a mirror can get to a refractor. And there are people who don't like the intrusion of prominent diffraction spikes in the image. Indeed, one could argue the same way for top-of the-range 5-inch apos and how they are easily bettered by a mediocre 8-inch Dob under good conditions. But that hasn't stopped 5-inch apos being built and used. Of course not!

Kerry Robbert from southwest Michigan purchased a DGM OA-4 and was asked about its pros and cons. "The 'scope eats power almost without limitation," he said. "My mirror is 97 mm (subtracting the 1-mm bevel from the diameter of the glass). Dan said they vary a bit. When these were introduced, they were popular as a "cheap" alternatives to apochromats, though with obviously superior color correction. At the time, nothing could come close for the cost. Now, however, the lowly Orion ED-100 apo refractor comes quite close to its performance (if you get a good one; and I did) and is slightly easier to mount than the DGM 'scope and arguably more versatile, too, owing to the ability to use large, 2-inch eyepieces on the former. My optical tube assembly – the version with the secondary on the same side as the focuser – suffers pretty badly from tube currents rising up off the ground. Because the light path is parallel to the top surface, more or less, depending on how the optical tube is rotated in the mount, it runs through the length of the current. Despite being a small mirror, thermal issues are quite prevalent. I'm still working on the best method to address the issue. Is an OA-4 worth it? I tend not to think so, given the quality available in f/9 Synta ED refracting telescopes. My Orion ED100 and Megrez 90 (apo refractors) are so close in performance and aperture, yet more versatile and easier to mount, that I rarely grab the OA-4 these days. Economics still favors the larger OA 'scopes over equivalent aperture apo's.

Fancy an 8-inch Planetary Dob?

If you think a 6-inch Dob is a little too small for your liking, then an 8-inch (200-mm) 'scope ought to show you significantly more. Sure, the larger aperture makes it more susceptible to seeing conditions and the larger mirror takes a bit longer to cool, but despite these misgivings, you'll get nearly twice the light-gathering power and a 33 percent increase in resolving power. The majority of 8-inch Dobs have focal ratios of 6 or lower, but some companies have seen a gap in the market for a larger aperture 'scope and longer focal length tailored to the needs of the lunar and planetary enthusiast. Ocean Side Photo and Telescope (OPT), based in California, have designed their 8-inch f/9 Planet Pro Dob with some very fetching attributes. The tube assembly is constructed from two thermally stable sonotube sections (upper and lower) that are held in place with four truss poles. The Lazy Susan base is made from Baltic birch, and comes with ebony star and Teflon bearings for smooth motions across the sky. The OPT 'scope also comes with adjustable side bearings to allow the balancing of heavy eyepieces and cameras.

The first thing you'll notice is how long this 'scope is. Fully 1.8 m long from top to toe, this 'scope goes on forever. A refractor that long would require a very beefy mount to do justice to the optics. With a Newtonian reflector, a simple alt-az mount – provided it's well built – can be more than adequate. The OPT planet Pro Dob has a very small (15.6 percent) central obstruction to maximize image contrast so as to deliver images that are refractor-like. The mirror is guaranteed to be "diffraction limited," multicoated, and made from Pyrex glass sourced in the United States. The entire instrument is surprisingly light – just 52 pounds – when fully assembled in its cradle and can be easily and swiftly disassembled for transport in the back of a small car. Heck, it even comes with a light shroud, a 30-mm wide-field eyepiece, and a Telrad finder at no additional cost! At $999 plus shipping, is this 'scope too good to be true? (Fig. 5.13).

As a moderately keen (certainly not serious) planetary observer, this author was taken slightly by surprise by the truss tube configuration of this 'scope. A truss tube is more susceptible to local seeing than a solid tube. Plus, there would be some concern about the 'scope's ability to hold collimation as the tube is moved across the sky. As most serious planetary observers will tell you, a good tracking mount is all but essential for detailed planetary studies, since the object can be kept centered in the field of view. Of course, it could always be fitted onto a dedicated equatorial tracking platform (discussed later in the book) if need be. That said, at f/9, such an instrument will serve up a large, diffraction limited field so you can still manage to track manually by employing wide-angle eyepieces and "nudging" the 'scope along. If the bearings are smooth enough, you can employ high powers to good effect.

Chris Hendren from OPT was apprised of these concerns and responded as follows: "We introduced the OPT ProPlanet Dob with the idea of bringing back the classic long focus planetary Newtonian in a more modern package that was easier to transport in small cars than the older solid tube designs. We are using an American-made 8-inch f//9 primary mirror and very small 1.3-inch secondary from Newport Glass

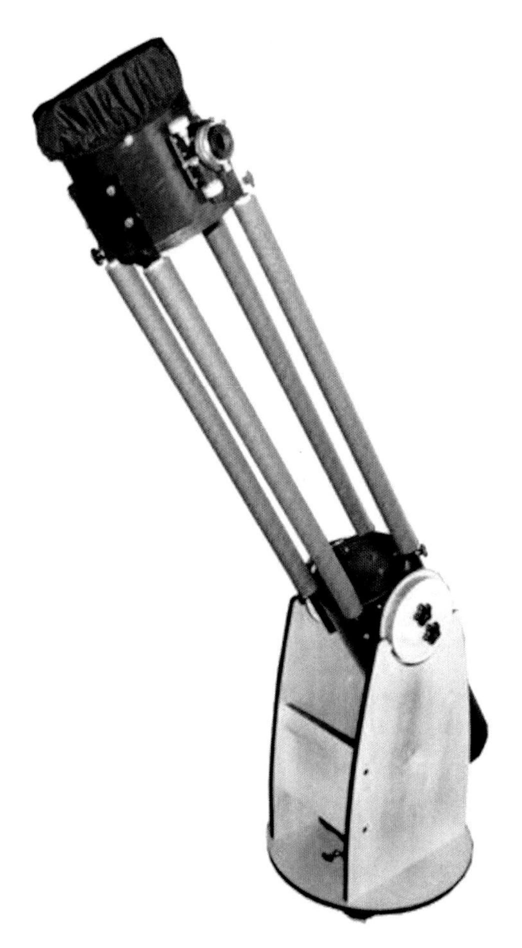

Fig. 5.13. Planet killer? The OPT 8-inch f/9 truss tube Dob (Image credit: OPT).

with the rest of the 'scope's structure and a curved spider being built by Discovery.

The truss design was an obvious choice in an era where almost everyone has to drive for good skies. Obviously planetary viewing can be done at home for many people, but they still have to get it home – often in a small passenger car – and they have to get the 'scope out and away from the city for the best views of planetary nebulae and globular clusters that the 'scope also excels at. A solid tube might be more thermally stable, but it also takes much longer to cool down, and there is no stray light or moisture issue as the 'scope comes with a fitted cloth light shroud.

I have seen my friend's 18-inch f/4.5 Obsession with a 7-inch off-axis mask outperform an APM/TMB 180-mm (7.1-inch) triplet apo on Saturn under the same conditions after a two-hour cool down, so the thermal issues are not really a problem for a standard truss 'scope. In this case, the lower 2.5 feet of the tube is sonotube, and the mirror cell is the same as a solid-tube Discovery Dobs, so ground thermals should not cause any more problems than they would for a standard Dobsonian.

Collimation is a bit of an issue for a truss 'scope, but with an f/9, the light cone is so forgiving that the alignment can be quite a ways off and still provide diffraction limited views. I have had hardly any issues at all maintaining collimation even with my 12-inch f/5 Dob through the night, and when we set up one of the ProPlanet Dobs at a star party last July with 300 people attending and dozens looking through the 'scope, the views were just as sharp at the end of the night as they were at the beginning – easily holding its own against 5-6-inch apo refractors on the Moon and Jupiter.

There is no off-the-shelf method of motorizing the 'scope, though it's possible that either JMI (TrainN'Track) or Tech2000 (DobDriver) would be able to adapt one of their tracking systems. Likewise, any cooling fan would have to be adapted by the owner. Orion sells a DC battery-powered fan for Newtonians for about $25 that could easily be adapted using Velcro or another simple attachment method.

Chris was asked about how well the mirrors were figured on these 'scopes. "Nothing is explicitly stated by the optician regarding minimum quality, he told me, "but the sample I tested looked to be on par with the star test as my 1/14 wavefront PV Raycraft refigured 12-inch f/5 mirror in my personal Dobsonian. At f/9 it is much easier to figure an excellent optic than at f/5, so I would guess that an American optician would be able to turn out fairly consistent 1/10+ wavefront PV optics, especially with Chinese and Taiwanese Dob makers regularly putting out 1/6–1/8 wavefront PV primary mirrors. My 'scope's mirror was originally made by GSO, and Alan Raycraft tested it at 1/7 wavefront PV before refiguring. I don't believe there is any published test data on the optics, though."

The Discovery Compromise 'Scope

Discovery optics has a reputation for being a couple of notches above the mass-produced imports, while not being quite in the same league with premium mirrors from companies such as Zambuto, Royce, OMI, and Galaxy. The construction is likewise a bit better than those offered by Orion,

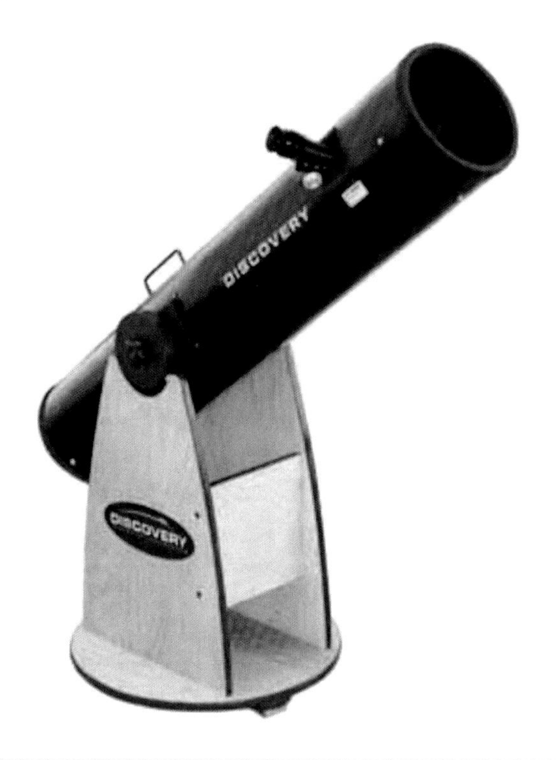

Fig. 5.14. The Discovery 8-inch f/7 PDHQ (Image credit: OPT).

SkyWatcher, and Zhumell. The company produces an 8-inch f/7 Dobsonian ($999) as part of their PDHQ range. Weighing a total of 41 pounds (including the mount), each instrument comes with a stylish 2-inch Crayford focuser with 1.25-inch adapter, a Telrad viewfinder, and 25-mm Plossl eyepiece to get you started (Fig. 5.14).

This telescope is very well built using durable, high quality materials. Optically, owners almost invariably speak of the precision optics at its heart. The mirror is very well figured, well above the diffraction limit, and comes with enhanced coatings as standard. Because of its shorter focal length, a given eyepiece will provide a wider field of view than it would with the 8-inch f/9 OPT Dob discussed above. So, it's less of a specialist 'scope than the others discussed in this chapter, but at f/7 it should still deliver killer views of the Moon, planets, and double stars. And at $999 plus shipping, it is a good deal on today's market.

In the next chapter, we'll delve into the brave new world of large aperture commercial Dobs, the light buckets that provide enormous light-gathering power at a price that won't break the bank.

CHAPTER SIX

The 12- to 16-inch Dobs

When making the transition from an 8-inch to a 10-inch or 12-inch aperture, portability is greatly sacrificed, but with the benefit of gaining more resolution and light-gathering power. The gain in brightness is always more profound than the gain in resolution. If you step away from an 8-inch Dob and place your eye at the focus of a 16-inch instrument, the first thing you'll notice is how much more bright (4-fold) the objects have become. Only afterwards will you notice the smaller gain in resolution (2-fold). Deep sky objects that are mere suggestions in the 8-inch 'scope come alive in the larger 'scopes, allowing far greater detail to be discerned to the careful visual observer. The greater resolving powers make these 'scopes killer performers on planets when conditions are right. There is a more dramatic difference in stepping up to a 10-inch Dob from an 8-inch model, compared with going to a 12-inch aperture from a 10-inch instrument. And it goes without saying that the move from 12 to 16 inches is sure to be immediately obvious to the eye. This chapter takes a look at the largest instruments commonly used by amateurs – those in the 12- to 16-inch aperture class. They offer huge light grasp and, when conditions allow, reveal the wonders of the universe in bright and stunning detail.

N. English, *Choosing and Using a Dobsonian Telescope*, Patrick Moore's Practical Astronomy Series 1, DOI 10.1007/978-1-4419-8786-0_6,

Large Aperture Dobs from Sky Watcher

In the previous chapter, we took a look at the 8-inch (20-cm) f/6 Skyliner Flextube Dob from Sky Watcher and its many innovative features, especially when you consider the price you pay. The largest in the current line of Sky Watcher Flex Tube Dobs is the 12 f/5. The instrument arrives in two boxes, one containing the optical tube and accessories and the other holding the flat-packed components of the mount. Assembly takes a little bit more than an hour. It is easy to put together thanks to the clear instructions provided. Once assembled, the tube, which weighs 45 pounds, is surprisingly easy to carry. Although more than 1.5 m long when fully extended, it's just 0.9 m long when tucked away collapsed, making it far easier to lug around in comparison to more conventional, closed-tube designs (Fig. 6.1).

The telescope's struts are attached to the upper-tube assembly, and when the tube is "collapsed" the struts simply slide through cylindrical bores in the lower segment of the 'scope. Its collapsibility is a FANTASTIC idea! In so doing, it combines serious aperture but in a tube length that can fit comfortably into the back of any vehicle. Three wing screws help lock the struts in place, and moving the upper-tube assembly is easy, but a bit firm. The tube is anchored to the base by screwing in two handles, one of which also functions as a tension-control adjustment. Collimation of the secondary mirror is achieved using a supplied hex wrench. And therein lies the first little niggle with this 'scope. These screws to have too much play in them – not to perhaps misalign the optics enough to prevent low and medium powers perhaps, but certainly enough to take an edge off high-resolution images of planets. That said, the collimation holds up very well between the "extended" and "collapsed" positions, so it's thankfully not something to worry about very often.

The upper tube assembly (UTA) itself seems a bit underbuilt. This was probably done to minimize the travel distance from the extended to the collapsed position. Further, the secondary mirror does appear to be a little too exposed. Because the UTA is only about 8 inches long, this is a 'scope that would benefit from a slight shroud to keep the dew off the secondary.

The focuser, too, is problematic. One thing it has going for it is that it's a Crayford. The focuser movements were very smooth and backlash free. Bizarrely, an adapter has to be used for both 2-inch and 1.25-inch eyepieces. The 2-inch eyepiece adapter has a flange that inserts into

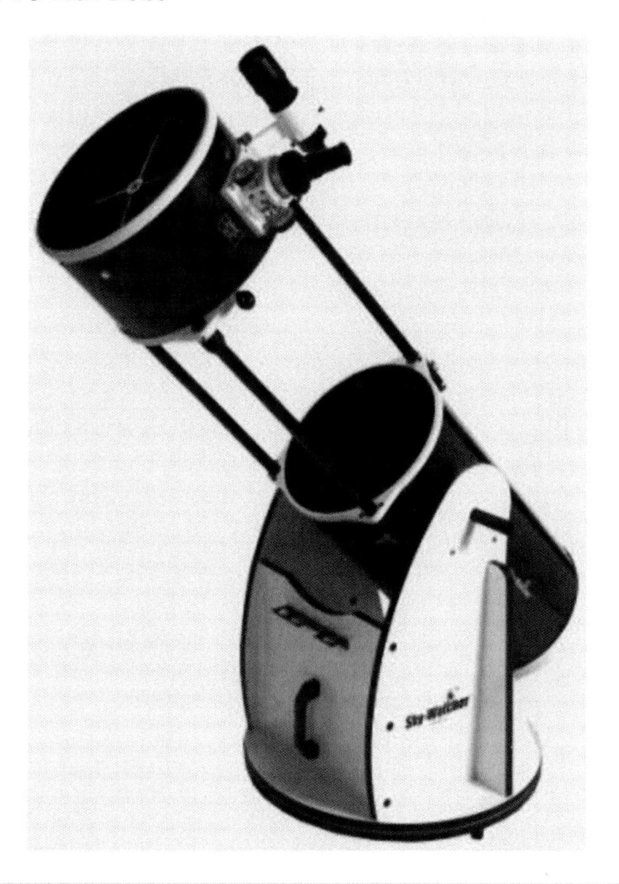

Fig. 6.1. The Sky Watcher Skyliner 300P Flextube (Image credit: Harrison Telescopes).

the focuser. This is perplexing, since most focusers are designed to accept 2-inch eyepieces from the start.

Nor is the mirror-reversed 8 × 50 right-angle finder 'scope supplied with the 'scope acceptable. Not that it is bad optically. On the contrary, it served up a wide and crisp field of view like any good finder ought to. A straight-through (but upside-down) finder would have worked much better, since it'd simply be a matter of rotating your sky charts upside-down to get your bearings. If you have a correct-image right-angle 8 × 50 Orion viewfinder you can simply swap them, since the difficulties of using a mirror-reversed finder 'scope wouldn't significantly hinder a rich-field, mirror-reversed refractor.

So, let's get to the meat of this beast. How does it perform on the sky? Well, collimation was more or less spot-on straight out of the box. Only minor tweaking needed. For sure, the length of the tube makes this a two-person job if you're doing it at night. And, as others have reported, it keeps its collimation time and time again. However, a fan is an absolute necessity with this sized 'scope, unless you store it out of doors. Surprisingly, this instrument didn't have one.

How does Saturn look through this telescope? With its rings nearly edge-on as it passed opposition, it was incredibly bright and sharp. That first look at the focused image confirmed what was seen in the earlier star test – this big Sky Watcher has a very good mirror. Good optical figure, together with its magnificent resolving power, are a killer combination when conditions are right. The Ringed Planet floated serenely across the field of view. In moments of good seeing, several bands could be delineated crossing the planetary globe, and the Cassini division popped in and out of view at the ansae. With the return of darkness in the early autumn, the Flextube 300 was turned on two globular clusters, M13 and M92 in Hercules. This 'scope really opens up these distant "satellite towns" of the Milky Way.

Although they lie some 30,000 light years away in the galactic halo, this big Dob opens these star clusters in ways that most other backyard 'scopes can't. The views of these clusters through this surprisingly portable large aperture 'scope are simply amazing. At 150×, the view is enough to leave you breathless – myriad stars filling the field of view and slowly concentrating towards a fully resolved core. On double stars it fares less well; the seeing rarely if ever permits one to exploit the full resolving power of this telescope. On balance, the 12-inch Sky Watcher Flextube Dobsonian ($995 plus shipping) is a very good value. The combination of decent aperture, ridiculously quick set-up, and crisp views come at a very affordable price. For those interested in owning a more conventional, closed tube design, the older SkyWatcher Skyliner 300P is still available at a significantly lower price.

A Big Dob That Finds Things

Orion (USA) Telescopes have marketed a large 12-inch truss tube Dob that finds things for you. Called the XX12i ($1,400), the package consists of a complete telescope, two eyepieces (10 mm and 35 mm), a cooling fan, an "intelliscope" hand controller, a collimation cap, and a copy of *Starry Night* software with DVD theatre. Assembly couldn't be easier.

You slot the 'scope into the altitude bearing. Two knobs are used to lock the encoder in place and provide adjustable tension. The truss tubes are fixed, two at a time, to the tops of the brackets and screw into substantial castings at the top of the lower section of the tube. With a big mirror housed inside a 9-point floatation cell, the base of the tube is quite heavy. The fully assembled 'scope and base together weigh in at nearly 85 pounds, so this is not exactly a grab and go 'scope (Fig. 6.2).

Collimation was quite good out of the box. Even after assembling and dismantling several times in succession, collimation required only minor tweaking. After cooling the 'scope for an hour outside with the supplied fan on, star testing revealed good optics. There was a trace

Fig. 6.2. The Orion (USA) XXTi (Image credit: Altair Astro).

of astigmatism at high powers, and overall it was a notch below the Sky Watcher mirror described above. Still, it seemed decent enough to serve as a good all-around 'scope. Like most eyepieces that come with these Dobs, upgrading them is advisable. Still, you will get one of the best views of M42, the Orion Nebula, with a telescope of this size, as described below.

Like some kind of bioluminescent creature from the depths, this giant molecular cloud showed up as an enormous, swirling mass of gas and dust lit up from the inside by neonatal stars. Vivid greens and subtle shades of blue and red could be made out along the border and the many filaments within. At the cloud's heart lies the beautiful quartet of newborn suns – the celebrated Trapezium. And on steady nights, the more elusive E and F components are readily discerned. This is what a telescope like the Orion XX12i was made for. Orion has also introduced an even larger member of its truss tube Dobs – the XX14i ($1,800 plus shipping and crating) for even brighter views.

There have been some reports in the literature that the quality of the mirrors these scopes are outfitted with exhibit some variability, ranging from downright unacceptable to very good. If you intend buying one of these secondhand, try to make some arrangement to test it out before you part with your cash.

If you prefer the closed tube design and are after an even better bargain, then why not consider the 12-inch Revelation "premium" Dob. Andy Sheen from Staverton, Wiltshire, gave us his take on this instrument. The "beast" comes in two boxes – the main box, containing the OTA, and a second flatpack box containing the base. These things are relatively big; the OTA box is about 1.7 m long, and weighs 24 kg, but still, if you want a big Dob, that's what you expect. Opening the boxes shows the innards. Firstly, the OTA. For something so large, it is rather lightly packed. (One OTA had to be returned, as the plastic packing ties had been used as handles by the couriers and gone through the cardboard and bent the OTA!) In comparison to Celestron – where things always came in two cardboard boxes – it is liable to damage in transit. You can also see the accessories that come with the OTA in situ (Fig. 6.3). This is what you get:

- 30-mm WA Revelation eyepiece
- 9-mm Revelation Plossl
- 35-mm extension tube (needed to reach focus on some eyepieces)
- 8 × 50 finderscope and bracket
- Battery cage for the fan in the base

The other box contains the mount and all the screws, as well as the accessory tray and some instructions on how to assemble from Telescope

Fig. 6.3. The goodies that come with the Revelation 12-inch Dob (Image credit: Andy Sheen).

House. The instructions were adequate, but they were for an older version of the mount, as there were some minor discrepancies between this mount build and that in the instructions. With a little careful thought figuring out which way is the bottom on the baseplate – it's actually with the screw insert at the bottom, you can fix the three stubby legs and have the whole mount built in about 20 min. Build quality is OK, but then this isn't a premium mount and it does rather look like it is from the Ikea Darth Vader collection (Fig. 6.4).

That said, it is functional. One point to note is to ensure you get the sides the right way around. The OTA uses some hefty machined knobs to mount into the rocker box. These are attached to the side of the tube via Allen bolts in a nifty sliding holder that allows you to alter the rotational point of the tube. Put the knobs higher, the tube becomes base-heavy, lower and it becomes top-heavy. Quite neat! Unfortunately, as these are fixed with Allen bolts, you can't change the balance in the field to make adjustments for the weight of different eyepieces (Fig. 6.5).

What you can alter is the tension in the pivot. Screwing the two knurled knobs up increases the turning resistance, while unscrewing them decreases the resistance.

Fig. 6.4. The completed Lazy Suzan base (Image credit: Andy Sheen).

Overall, for the price, the details of this OTA are impressive. First, the round circle appears centrally placed and the mirror mount looks sturdy and has a fan to cool it. Additionally, the collimation/locking knobs seem of good quality although you may need to change the springs for some from Bobs Knobs. Conveniently, the underside of the mirror has a built-in cooling fan that operates off a 9 V battery. Left outside for an hour cooling without a fan and it still isn't ready. Twenty or so minutes with the fan on and it's fine (Fig. 6.6).

Another thing that impresses about this 'scope is the focuser. It is a dual-speed Crayford and appears to be very smooth (Fig. 6.7).

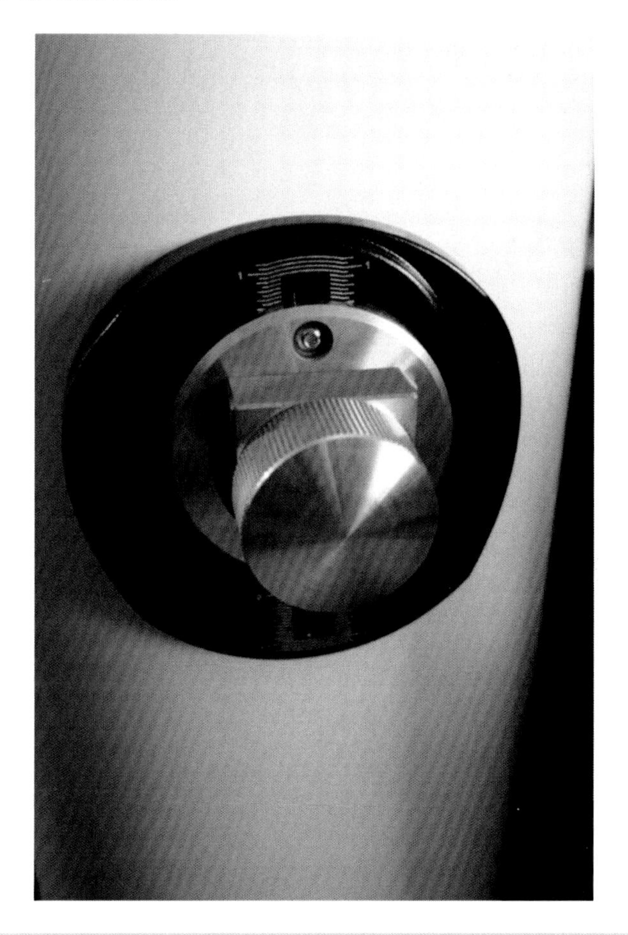

Fig. 6.5. The large knobs used to affix the OTA to the mount (Image credit: Andy Sheen).

Let's compare the Revelation 12-inch Dob with the C9.25 Schmidt Cassegrain telescope (SCT) after having been out in the garden for nearly two hours before beginning observations. First thing is to check collimation. The C9.25 requires a small tweak while the Dob didn't need anything, as it had been collimated earlier with the Barlowed laser method and checked with a star test. The Milky Way could be seen clearly through Cygnus and the Andromeda Galaxy and Double Cluster could be easily picked out with the naked eye. So, with everything collimated (and the C9.25 aligned) the first targets were deep sky objects.

Fig. 6.6. The completed tube and mount (Image credit: Andy Sheen).

First up was M13, the great globular cluster in Hercules. Here the Dob seemed to show more detail and was able to resolve stars to the core more easily. The SCT was murkier looking by comparison, and there wasn't as much "pop" to the image. Both 'scopes gave a good view, but the Dob was definitely better. Now onto the planetary nebula M57 in Lyra. Again, the Dob seemed to show a brighter image, although both 'scopes had enough detail to be able to see the ring, the darker center, and a hint of something lighter again right at the center. Now the famous double double (epsilon 1 and epsilon 2 Lyrae). The result was about a draw. Both 'scopes resolved the binary stars well.

Fig. 6.7. The quality Crayford focuser on the Revelation 12-inch Dob (Image credit: Andy Sheen).

How about Jupiter? By now, the giant planet was appearing over a neighbor's house. Both instruments gave an excellent view of Jupiter, although again, the Dob edged it – not quite on scale, as the C9.25 has a longer focal length, but more detail seemed apparent in the Dob. The Great Red Spot could just been seen through the Revelation Dob, but it was harder to make out in the SCT.

Now to the galaxy pair M81 and M82: With the longer focal length eyepieces, you could get both of these in the viewport of the Dob at the same time. Needless to say the views were lovely. They were easily bright enough to delineate some structure. In the C9.25, both galaxies were there, but again, the darker images it presented just meant they didn't stand out as well.

What about the Perseus double cluster (NGC 884/869)? This took quite a bit of time, but it seemed a good test of stellar resolution. Although the C9.25 was a tad darker again, all of the stars were visible in both 'scopes. Other objects examined in the comparison included the fainter planets Neptune (just) and Uranus, M103, and a few other galaxies. All consistently showed a brighter and "better" image in the Dob than in the C9.25.

What's more, the Dob is far easier to set up, and star hopping is relatively easy if you have enough stars visible and a star map. The C9.25 (on a CG5-GT) is noisy and slow to set up by comparison and requires star aligning – although the targets are tracked once this is done. However, nudging the Dob seems to be reasonably simple and can be gotten used to. In conclusion, as a visual 'scope, the 12-inch is really rather good. If one had to choose between the C9.25 and Dob for visual, no question, it would be the latter. The quick setup and better views in the big light bucket win every time.

The most amazing thing about the Revelation Dob is the price. The entire package can be had for just £500 complete.

The Orion Optics UK, OD 300 L

As previously mentioned, Orion Optics UK has a well respected reputation for producing finely crafted Dobs of all shapes and sizes. The OD 300 L (price for the standard ¼ wave optics package is £859 plus £308 for the mount) has a slightly longer focal length (f/5.3) than the standard, mass market Dobs in the same aperture class and can be purchased with deluxe optics and various other accessories at additional cost. Shane Farrell from Manchester in England provided his take on this big, British-made Dob. "This might seem a little steep for a used 12-inch Dob," he said, but I took the view that the quality of the base, the renowned optics, and the slightly longer focal length would mean that the 'scope would be a good buy and last me a long time. A nice surprise was that it came with a Telrad affixed to it, a friction brake, and also new caps for both ends. More importantly, the 'scope has been flocked internally for most of its length, has a cross-primary fan, and one of the thinner mirrors previously used by Orion Optics. This means that it cools down surprisingly quickly with the fan on, and I have run it constantly on the few occasions I have used it. There is another nice modification, which is a small drilled hole on the opposite side of the focuser that allows a pin to be inserted so that you can ensure the focuser is completely centrally mounted from time to time (Fig. 6.8).

"An addition made was a new "retaining ring," which is basically a single OTA ring fitter on the inside of the bottom of the two rings. This performs a couple of important functions; firstly, it allows the tube to be turned if required without it dropping down. This ability to turn

Fig. 6.8. The Orion Optics UK OD 300 L (Image credit: Orion Optics UK).

the tube (and indeed add tubes from other manufacturers of the same diameter) is another feature of the Orion Dob base – most other 'scope designs do not have the facility to turn the tube and therefore if you find the designed angle not to your liking you cannot do much about it. Secondly, it adds a little weight to the bottom half of the tube. This enables you to move the tube slightly upwards, which gives more flexibility when adding weight at the top end with large eyepieces and other items of equipment. My first impressions (having never actually seen a 12-inch Dob before I visited the seller) were "Wow, this is awesome." My opinion has not altered since.

The tube itself is surprisingly light. I am tall (6′ 3″) and reasonably strong but find it is easily lifted off the base and carried into the garden. For carrying purposes, the design is excellent (although this may be "accidental/incidental"). There is a cradle with rings and so-called trunnion blocks that allow the OTA to be picked up at the balance point by very sturdy fixings. I usually lift the tube off the base, stand it up, and after carrying the base outside, carry the OTA out and set it straight onto the mount. I then let the fan do its job, and within 45 min to an hour I can be viewing – not bad for such a large Dob from what I read. Actual set up time is about two minutes – i.e., the time it takes to carry the optical tube assembly and the base the length of the house. I was originally going to store it in the shed to reduce cooling times, but paranoia about theft, dirt, and bugs, and the fact it is so easy to carry and has a small footprint, as well as the quicker than expected cool down times from "warm," mean that I have no issues leaving it safe in the house.

As with most people, probably, one of my first targets was the Orion Nebula. This was really lovely though the viewfinder, and with a 14-mm Televue Radian eyepiece I was able to see E and F (just) in the Trapezium with direct vision. These are much more readily seen when using my 2.5× Televue Powermate. With my 18-mm Radian eyepiece, the nebula seems to go on forever, and there's a lot more detail than I have seen before. I also found another small patch of nebulosity lower down than M42 and am still trying to work out what this was. A word of warning when using the Powermate; it sticks out a fair bit with an eyepiece in and be careful not to smack the side of your head on it – this will do you or your equipment no good at all – yes, I am speaking from personal experience!

I then decided to try and split some double stars. Obviously most were easy targets, but Castor, Polaris, Rigel, Mizar and the two doubles in Orion's belt were very nice indeed. Double stars and planets are not generally what this 'scope is designed for, but being inexperienced, I find the faint fuzzies still hard to track down. Mars was excellent with far more detail than I have seen previously – the main ice cap very obvious as was a fair amount of greenish "mottling" on the main disc. Later in the night, Saturn eventually appeared. Through the OD 300 L the planet was stunning, with at least four moons visible as well as great detail on the rings; the Cassini division was obvious with the 18-mm Radian in the Powermate and even more so with the 14-mm Radian, although you lost a little definition on the planet disc at the higher magnification.

Some darker bands were noted on the Saturnian disc itself. I have found that the combination of a Telrad as well as a 90 degree 6 × 30 finder brilliant for locating targets quickly and with no fuss/less neck craning, even when using quite high magnifications when I cannot be bothered taking out the Powermate and 18-mm Radian, for example, to then use a "finder" eyepiece and then put everything back. This gives magnification of about 220×, and personally I am really impressed with this, as it makes the initial fine-tuning of the finders worthwhile.

The Dobsonian design really does what it says on the tin. It is just so easy to set up and use with intuitive pointing, and with the wider angle/field of view of the Radian series (60 degrees) it is very easy to keep targets in view even at high magnification. I find that getting the image on one side of the field and then letting it drift slowly across is the most enjoyable method. The ease with which things can be found and tracked was proven recently when I was just scanning about near Ursa Major with my 24-mm Panoptic (67×) when a small satellite (not visible to the naked eye) zipped across the view. I put the 'scope to where I thought it should be, and there it was. Despite it moving very quickly, I managed to track it for maybe 30 s or more until it went out of view. Based on this, I should be able to manually track objects at high magnification eventually. Even with a long 1,600-mm tube, a chair to observe is essential in my view, as I have to stoop to view even at the zenith. I have made a Denver Chair out of odds and sods. One day I'll get around to making an even better one, but it works a treat for now.

I like to fine tune the 'scope's collimation every time I observe and use the Barlowed laser method. This takes just a couple of minutes each time as it's generally quite accurate. I did some practicing on a smaller 6-inch Celestron Newtonian, which I sold in part to pay for the Dob, so hopefully I have it pretty well sorted now. Although I still really love my Celestron OMNI 120-mm refractor (which is also a "keeper"), I am now a big fan of Newtonians and Dobsonians in particular and Orion Optics UK instruments specifically. I'd strongly recommend anyone looking to get the most out of their available time to try and look through a large Dob. Their big aperture increases the resolution and allows things to be seen in much more detail than I have seen in smaller 'scopes however perfectly made they are. Though I still consider myself a beginner in this hobby I can genuinely urge anyone to not worry about collimation – this can be picked up quickly – or the size of these 'scopes. Just get the largest you can handle. For me, the views I have had in the short time I have had this 'scope have been better than anything I have tried previously. I think it was Mick (The Doc) who said that a large Dob and premium quality eyepieces is as good as visual observing gets, and I must admit, I tend to agree!

American Favorites

On the other side of the Atlantic, U.S.-based company Discovery Telescopes offer a range of instruments that apparently rate comparably with the best the Brits can offer. The company has taken a lot of flack on the Internet owing to their failure to adequately respond to customer inquiries and the very long waiting times buyers have had to endure. Indeed, in his book *Star Ware*, Phil Harrigton went so far as to not recommend the company, even though their products are of very good mechanical and optical quality. That said, there are evidently enough happy owners of these 'scopes out there to warrant a discussion on the Discovery Dobsonian product range. In particular, the company's PDHQ series has been lauded by amateurs up and down the country for delivering very good optics and first rate mechanicals to boot.

One enthusiastic owner of the 12.5-inch f/5 PDHQ ($1,699) said this: If you want great optics in a customizable package, this 'scope is for you. Be prepared to wait longer than the estimated delivery time (one month in the Continental USA). Mine took two months. For the most part, the packing was well done and protected the 'scope fine. Two minor nits were just a few crumbs of sand on the mirror, which were easily blown off, and some spray foam got on the side panel of the rocker box where the protective plastic pulled away. Assembly is very straightforward. The 'scope tube comes in two parts and fits together easily. I had a couple of the minor issues; the post for the finderscope bracket wasn't aligned to the optical axis, so the finderscope itself is skewed in the opposite direction to compensate. Secondly, the altitude pads were too small and too far apart to allow for smooth movement.

I picked up a larger piece of Teflon and cut it to the correct size as well as moved it in to more of a 90 degree position (with respect to the center of the trunnion). The 'scope now moves silky smooth. I've also added some gussets to stabilize the rocker box, removable wheelbarrow handles to cart it around, and installed some fans, but that's more of a personal preference thing. When the skies cooperate the Discovery throws up views about as good as a 12.5-inch reflector will allow. So if you are at all handy and willing to put in a little work you can save some quite a few dollars over the bigger names and have a 'scope that performs about as good.

You know what they say, don't you? You get what you pay for. There's certainly some truth in that statement, especially after you take a look at what companies such as Obsession and Starmaster are producing in the large aperture market. Obsession telescopes, based in Wisconsin, produce high quality truss tube Dobsonians ranging in size from 12.5 inches to 25-inch behemoths. Back in the 1980s, large Dobsonians were readily available, but their optics and mechanics were only so-so. Company founder Dave Kriege was one of the key individuals who changed that perception, by bringing high quality optics and innovative mechanical design to his Dobsonian range. Truss tube design, semicircular altitude bearings, and multi-point mirror cells with an open back are all innovations that Kriege brought to the market. Phil Harrington, in his book *Star Ware*, has gone so far as to suggest that these innovations were so far reaching that the new design should really be called "Kriegian" rather than "Dobsonian." Here we'll take a look at one of their most popular models, the 12.5 inch f/5 truss tube Dob.

Right off the bat, the Obsession 12.5-inch ($3,295 plus shipping and crating) is a very thoughtfully produced telescope. The primary and secondary mirrors are made by Galaxy Optics and Optical Mechanical Inc (OMI), respectively. The optical quality is guaranteed to be better than ¼ wave (peak to valley) across the entire face of the mirror and uses enhanced aluminum coatings reflecting back 96 percent of the light hitting it. The secondary mirror is a little over 2 inches along the minor axis, giving a 17 percent central obstruction for more contrasted views and has "super enhanced brilliant-diamond" coatings that deliver 98 percent reflectivity. The secondary is supported by curved spider vanes, so there's no diffraction spikes for more refractor-like images (Fig. 6.9).

What does that all that mean? Well, let's say you needn't worry about the optics. The whole 'scope is very easy to set up. The eight truss poles are inserted into wooden blocks mounted on the rim of the mirror box, and then the upper tube assembly is placed on top and secured with cam-levered aluminum clamps. It couldn't be simpler. Although it sounds less than ideal, owners report no problems holding collimation even after several long road trips. With its built-in 12 V dc fan, cool down is fairly quick and easy. Your extra money also pays for silky smooth bearings on both altitude and azimuth axes, making tracking a breeze, even at high power. Special glass and phenolic anti-friction laminates on all bearing surfaces glide across virgin Teflon, resulting in nearly equal forces in both altitude and azimuth. These so-called "Ebony Star" surfaces provide smooth motion from the tip of your finger. One additional bonus is that the telescope is fully compatible with the powerful Argo Navis digital setting circles; a real bonus when searching for faint fuzzies.

Fig. 6.9. The 12.5-inch Obsession Dob (Image credit: Obsession Telescopes).

A large Dobsonian vendor and keen amateur astronomer who wished to remain anonymous was asked for his perspective on where he felt Obsession telescopes scaled in the quality scheme of things. "The Obsessions are very good medium grade telescopes," he said. "Construction quality is excellent and optics are fine, and I'd certainly be happy with one of the Obsessions as a lifetime investment. That said, I've seen many 'scopes that were superior in construction quality and which had even better optics. Of course, these 'scopes were considerably more expensive than the Obsession range, so if cost were no object, you can get better. A good way to think about is that among the premium 'scope makers, Obsession is considered "the affordable" one. Some compromises in materials and workmanship are made to keep costs down.

Reaching the Stars With a Starmaster

Once you go beyond the 12-inch aperture class, you encounter a number of problems that are not easily surmounted if you still want to keep a 'scope portable. An 8-inch f/6 Dob is 48 inches long. A 16-inch f/6 would be twice as long again! Unless you're on anabolic steroids, you can see immediately that this 'scope is not the kind you can lug around your back garden easily. One way of getting around the problem is to shorten the focal ratio of the telescope. But faster mirrors, as we have seen, are more difficult to make well in comparison to their longer focal ratio counterparts – difficult yes, but hardly insurmountable. To see what we mean, you have to take a look at the Starmaster range of ultra fast (F < 4) Dobsonians.

When it comes to quality Dobsonians, few would argue that those produced by Rick Singmaster, founder of Starmaster Telescopes, Arcadia, Kansas, would not be near the top of the list. The company has gained a great reputation for delivering some mighty fine, large, ultra portable Dobsonians ranging in aperture from 11 up to 30 inches. In the smaller apertures, at least, Singmaster used to offer higher focal ratio mirrors, but these days he has clearly concentrated on producing quality optics faster than f/4 for uncompromised portability and optical quality.

Starmaster Dobs come in two varieties: the FX series, which includes instruments ranging in aperture from 14.5 up to 30 inches (with focal ratios of either f/4 or f/3.7) and the super FX series, which includes telescopes from 20 to 30 inches in aperture but are offered at an amazing f/3.3! Later we'll take a look at Singmaster's bigger 'scopes. Here, we'll talk about the Starmaster 16.5-in. f/ 3.7 FX Dob.

Singmaster employs oak plywood for his primary components. Contrast that to other manufacturers, which use apple ply composed of thinner layers of laminated veneers. Oak plywood is stronger, heavier, and more expensive than the others, but it also means the instrument is more durable. The rocker box and mirror cage are both constructed from this material. The Starmaster 'scopes all utilize a light stain that complements the natural grains of the oak. The finish Singmaster applies is thicker and gives the construction a fine glossy furniture-like appearance. The mirrors of all the FX 'scopes are mounted in a sturdy metal cell that can be removed from the mirror box between observing programs. The upper cage assembly is constructed of two plywood rings separated by a short length of sonotube, which is attractively finished in matte black.

The standard focuser is a very nice Crayford design, although it was a bit surprising to hear that it is of the single speed variety. Such fast optics seem to cry out for a dual speed focuser. Starmaster does supply an upgrade, but it's an additional $325 for the privilege. The trusses come in pairs, configured in a triangular assembly. Each pair is attached on either side of the mirror box and then affixed to the upper cage assembly. The trusses are constructed of hardwood dowels. Set up is very easy and, as a one-person job, takes no more than five minutes. Each of the lightweight plywood boxes can be easily handled and put together. Once the 'scope is assembled, the tube is tipped over horizontally to allow the mirror cell to be slipped into place.

One of the nicer features about the Starmaster is that the entire lower tube assembly – including the rocker and mirror box – and the UTA cage assembly stack together, one inside the other. The optics are almost unanimously agreed on by amateurs who use them to be first rate. With mirrors supplied by Lockwood Custom Optics, you're assured of getting great images. Indeed, Rick inspects each 'scope under the stars before he ships it to assure his brand's reputation. Singmaster also spends time in the field with his customers to share their joys as well as learn of any problems they encounter.

The Devil in the Details

The differences between the functionality of the primo truss tubes are subtle at best. For example, some observers have noticed that the action on the Starmaster line is stiffer than that found experienced with Obsession 'scopes. What are we to make of that?

It's all about the fine art of "nudging," which we shall loosely define as the knack of moving a Dob manually across the sky as smoothly as possible, either at low or high power. This is, after all, the most "organic" of ways to explore the starry heavens. Although you can nudge the Obsession telescope with one finger control, you pay the price when you use heavy eyepieces.

With a stiffer action, the total balance of the 'scope is not such an issue. You can go from a heavy, low power, 2-inch eyepiece to a lightweight high power ocular and the 'scope won't move when you don't want it to. Another nice additional option you can get is the GoTo system. Up to fairly recently, Starmaster was the only company offering this option for its customers. Consensus appears to report that it performs flawlessly.

Fig. 6.10. Rick sitting alongside the Starmaster 16.5-inch FX dob (Image credit: Rick Singmaster).

The motors are quiet and make a delightful little chirping sound at the end of a slew. Singmaster makes some mighty fine telescopes. Choose the one that suits your lifestyle and your wallet (Fig. 6.10).

Teeter Triumphs

Teeter Telescopes (founded 2002), based in Lake Hiawatha, New Jersey, USA, has built a sterling reputation for designing exceptionally high quality truss tube Dobs finished in wood. Offering instruments that range in size from 8″ through 20″, everything's built in, including state-of-the-art cooling fans, a high quality, dual-speed microfocuser and exceptionally rigid truss tubes that hold collimation under conditions where other brands let you down. Customers can also choose the source of their first class optics. What is especially admirable about proprietor Rob Teeter's range is its flexibility. While some customers seek ultrafast sub F/4 instruments, especially in the larger apertures, others prefer the classic F/6 variety and still others opt for more specialised scopes, such as their F/8 "Planet Killers."

Now it's time for something completely different. We take a closer into the heart and soul of the Dobsonian movement and explore how the "simplicity" of observing that John Dobson espoused in his dialog with the amateur community has been "re-invented" in a ways that transcend the norm. Time to see different strokes of the Dobsonian movement.

CHAPTER SEVEN

Different Strokes

If we are to interpret the Dobsonian revolution in the broadest sense of the word it is the enjoyment of a reflecting telescope free from the hassle of complicated, and temperamental, electronic mounts. The majority of Dobs thus far discussed are mounted in a rocker box-style mount called a Lazy Susan or affixed to a single axis arm (such as smaller Orion Starblast 'scopes discussed previously). In this chapter, we'll take a look at Newtonian reflectors mounted on unusual alt-az mounts that look and feel entirely different from the ones discussed so far.

First up is a modern classic that has been revamped for the twenty-first century. Everyone remembers Edmund Scientific's little Astroscan. Launched in the mid 1970s, this cute little 4-inch 'scope, finished in brazen red, brought extreme portability and decent light-gathering power to the masses at an affordable price. The instrument, constructed from an extremely durable plastic, sat on a little ball and socket mount that could be quickly positioned to point at any object in the sky. That much hasn't changed in the new and improved Astroscan Plus. At its heart is a 4.1-inch f/4.3 parabolic mirror. What's more, the Scientific's website states that this mirror has a figure of 1/8 wave and so theoretically should be capable of pretty decent low and high power images, despite its 36 percent central obstruction. In practice, though, one problem with attaining good high powers with this telescope is that the optics are sealed off behind a clear glass window in order to avoid children tampering with the optics.

N. English, *Choosing and Using a Dobsonian Telescope*, Patrick Moore's Practical Astronomy Series 1, DOI 10.1007/978-1-4419-8786-0_7,

Astroscan Plus comes with two upgraded eyepieces – a 28-mm Plossl yielding 15× and a 15 mm Plossl serving up a power of 30×. These are a noticeable improvement on the original Kellner eyepieces that came with the earlier versions of the 'scope. The focuser has also been fitted with Teflon bearings to improve its movement and is also supplied with a unit power red dot finder. Weighing a mere 13 pounds including the base, the 'scope can literally be taken anywhere at a moment's notice. For wide-field vistas in a hurry, this 'scope is hard to beat. With the supplied low power eyepiece, you get a nice 3-degree field enabling you to frame lots of your favorite deep sky objects easily (Fig. 7.1).

A review of the new and improved Astroscan Plus, conducted by Gary Seronik in the July 2010 issue of *Sky & Telescope* magazine, revealed that the 'scope worked well for low power, wide-field sweeping, but images became "noticeably soft" at powers above 50×. He noted that the most

Fig. 7.1. The newly upgraded Edmund Scientific Astroscan Plus (Image credit: Scientifics).

likely cause of this compromised performance at high powers was a slightly mis-collimated optic. That said, the Edmund Scientific Astroscan Plus comes with a 5-year warranty (like the original), so any performance issues, should they arise, can be resolved quickly. The Astroscan Plus can be purchased as a standard ($229 plus shipping) or deluxe package ($429 plus shipping). The standard package gives you everything described above, as well a slip-fit dew cap, an up-to-date 36-page user's guide, an adjustable shoulder strap, a 35-page "Sky Guide" Booklet, and a star and planet locator. The deluxe package also includes a roof prism, which orients objects right-side up and left-to-right, a Sun-viewing screen, a 2.5× Barlow lens, a nylon tote bag, and a lens cleaning kit.

Let's now look at a couple of telescopes manufactured by Parks Optical. Known as the Astrolight systems, the company produce two instruments, a 6-inch f/6 ($799) and an 8-inch f/6 ($999). Both instruments come with very unusual alt-az mounts. Standing 45 inches high and weighing just 25 pounds, both 'scopes are mounted on a smooth alt-azimuth pier system made from an advanced glass-filled high impact composite ABS material. The 'scopes come with 1.25 rack and pinion focusers, with no facility to use 2-inch eyepieces. This is probably because the secondary is too small to fully illuminate the large field in a typical low power, wide-angle eyepiece. Another surprise is you get a 6×30 finder, adequate certainly, but nothing to write home about. Some observers also quip that the central pier is too tall to accommodate most seated observers. Perhaps the biggest surprise of all is that the primary mirror is not center marked, which might present real problems when it comes to collimating the 'scope.

Kerry Robbert from Michigan provided a forensic analysis of his experiences with the Parks Astrolite 6. "The fork mount required some file work to allow the altitude trunnions to fit through the tops of the notches. It's not a big deal, but it's irritating. The tube rings had injection mold "nipples" on them that would catch on the bottom of the OTA as you tried to rotate it for both eyepiece location and balance. I filed these off as well. Again, it was easy to do, but irritating. The altitude bearings are plastic on plastic. The bearings are approximately 1-inch diameter cylinders that ride in 1-inch diameter slots in the fork. They work surprisingly well.

The friction plates – two large threaded knobs that pinch the fork together increasing friction – work well. The azimuth bearing is two sets of thrust bearings, one set above the pivot bolt, one set below. There's a nylon-tipped nob/bolt that can be adjusted to press on the top surface of the azimuth plate, for increasing azimuth friction. This works well. The three legs each have a bolt on the lower side, which pass through holes in the steel pier and are attached with nuts inside (I replaced them with butterfly nuts for no-tool setup). The tops of each leg are clamped to the pier with a hinged plastic ring, similar to an OTA tube ring. Each leg has a leveling adjuster. The only metal parts include the bolts, pier, azimuth bearings, and azimuth surfaces. Everything else is plastic, and they feel like plastic. The OTA/mount clicks and creaks like plastic. Though maybe not aesthetically pleasing, it is nonetheless very functional. The mount is quite wiggly – dampening times after a tap are 7–8 s – and the induced oscillations make focusing a little tedious. Dampening times can be reduced a little by careful movements. Vibration suppression pads pretty much fix the issue. The legs are a bit tedious to remove. Their spread is very large. Getting the assembled mount out the door is very similar to moving a table – you try to weasel one leg out at a time, trying not to clunk the door and trim while doing so.

This 'scope is decidedly not "grab and go," and I find it best to leave it in the garage – removing the legs is a bit tedious. The fiberglass tube has a thickness slightly more than 2 mm and gloss white impregnated on the outside, with curly fiber matting inside. The inside of the tube is painted matte black and does the same job as flocking material, though not quite as dark. The focuser is an all-metal 1.25-inch no frills affair. I find it's a touch stiff. The finder is nicely finished, too. Oddly, it's mounted well behind the focuser. I prefer them further forward to make them easier to get under. It's also mounted on the left side of the tube assembly when viewed from the rear of the tube, and so it's better for left-eyed people. I'm right-eyed, so I have to position myself so that the finder is between me and the focuser. I'll likely move this to a preferred position.

The tube ends are capped by plastic tube rings, friction fit for easy removal. The front one arrived cracked. I'll attempt to get a replacement. The cell is all plastic and very spindly/airy for good ventilation. It fits in the tube nicely and was easy to remove for center-spotting. The mirror is glued to the front portion of the cell, in a recessed carrier. I have no idea how one would get at the glue for cutting when the mirror needed re-coating. Rather than three globs of glue evenly spaced, the entire rear of the mirror is glued nearly flush to the surface, wherever it contacts the spindly/airy cell.

Collimation movements are smooth. The spider is of the classic thin-metal band four-blade variety. The secondary is glued to the end of the stalk, collimation is by three Allen head screws. Works well. The fiberglass tube works wonderfully well for suppressing thermal currents. One of my pet peeves is steel-tubed Newtonians, which radiate heat quickly from the top, and pick it up quickly from below, creating an unwinnable thermal battle in the light path.

The mirror cools surprisingly slowly, considering the open nature of the cell. I suspect that if the mirror were lifted off the rear of the cell, rather than glued flush, cooling might be a bit faster. The mirror shows significant thermal deformation for about 45 min under fairly mild temperature change. Stray light control is good – the focuser is set far enough back, and the focuser tall enough so that stray light can't easily reach the focal plane. The tube is very dark inside, so deep sky object views were sharp and very contrasty. Once cooled, the star test showed a very nicely figured paraboloid. But once things really settled down I noticed the high-power Airy disk was really tiny – too tiny. Then, once things were settled enough for me to look at the surrounding diffraction rings, I was able to see that they were broken in three evenly displaced locations. Gently racking back and forth through focus gradually revealed the culprit: a pinched primary.

While the pinch is mild, it had a pronounced effect on resolution above 150×. This wouldn't be a big deal, but, as I mentioned earlier, getting the mirror out of the cell will be difficult – there's not even room, I don't think, to get dental floss in there for cutting the silicone. I called Parks to find out what to do and they instructed me on how to relieve the tension on the primary a little bit, which rectified the problem.

When I had the cell apart to look for possible stress points, I noticed the silicone was fairly brittle – probably the water-based version, or something like that. In any case, I was able, with moderate force, to push the mirror free (no space to get anything under the mirror, dental floss or otherwise, so cutting was not an option).

Re-mounting the mirror with double-stick foam tape, backed up with duct tape on the mirror edges (a mounting method I learned from taking my DGM OA-4 cell apart), solved the problem. Dampening times went down markedly when I removed the large plastic disk feet from the bottom of the levelers and replaced them with rubber slip-on feet from the hardware store. I observe on concrete, and I suspect the wide disks were sitting a little on their edges, and so would flex along their lengths (probably intended for grass surfaces, for which they'd probably be fine). I can't come up with another hypothesis why dampening times would suddenly drop. My intent with removing the large disks was simply to get the feet to fit more easily on vibration suppression pads, but the suppression pads are no longer necessary. Dampening times are down to 3–4 s after a good tap, 2 s when guiding or focusing gently – quite good, and more in line with what I'd expect given the size of the mount.

The duct tape is only a backup to keep the mirror from falling out of the cell should the foam tape fail. It's three ¾-inch wide sections at 120 degrees – not enough to reduce airflow or insulate, I don't think.

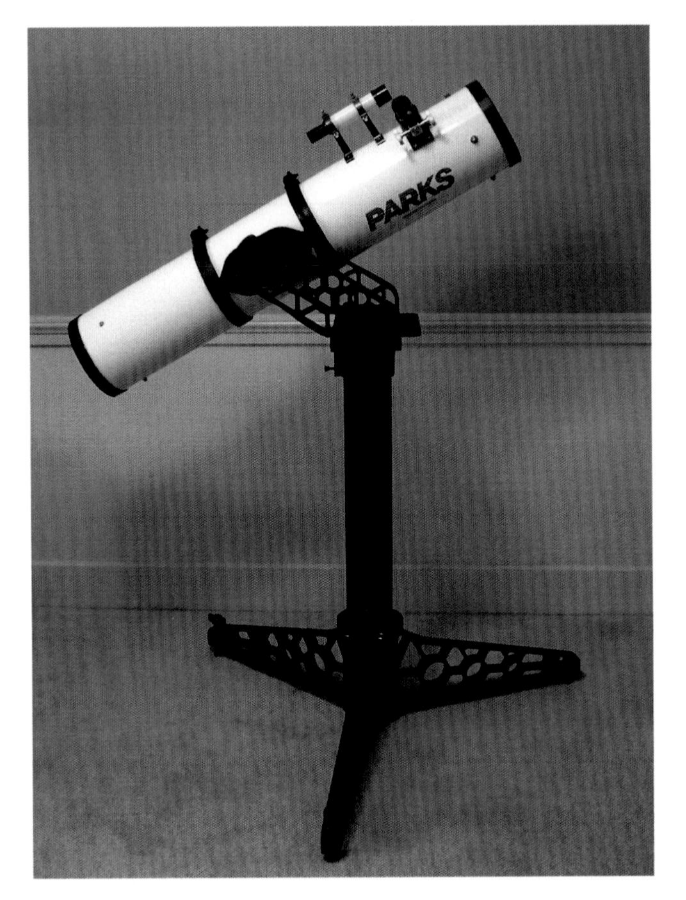

Fig. 7.2. The Parks AZ-6 (Image credit: Kerry Robbert).

I center-marked the mirror myself with a "cats eye" center mark and template. I actually like observing standing, so I doubt I'll cut the pier. The fact that the mount was designed for standing use was one of the many things that drew me towards it (Fig. 7.2).

The optics are now absolutely superb, and there was no need to send the cell back to Parks. I put the 'scope through a thorough testing over several nights. Saturn, some doubles (fairly poor seeing), lots of Messier hopping in Ophiuchus and Canes Venatici, along with several "standards." Everything suddenly seems to be working remarkably well – quite different from my first night out. I wonder how many of these 'scopes are suffering from this problem. It's unfortunate, because the 'scope is largely aimed at beginners, who would likely not recognize the mild

warp and its symptoms, or recognize the severity of its impact on resolution. Such a person would likely later simply decide Parks optics just aren't that good, after, say, having looked through a good sample of an Orion XT 6 or a good C6-R. Parks may be shooting themselves in the foot, which I hate to see, given how good (excellent is a better word) their mirrors and tubes are.

Overall (excluding the pinched mirror) I'm very pleased with the 'scope. But, it wasn't the "oh so cool" experience I was hoping for. The creaks and groans of the plastic, the protracted cool down, the focus wiggles, the finder location, the stiff focuser – all conspired to dampen the experience.

This is a decent and functional 'scope and mount, but it's by no means awesome. Once the mirror was "fixed," the views were flawless. I prefer this optical tube to the steel-tubed varieties. If only Parks did this OTA (or the f/8 version) on a conventional Dob mount; it'd be easier to get out the door in one trip and would likely be more stable to boot. The only reasons to choose this 'scope over, say, an Orion XT6 Classic would be: fiberglass tube (a big deal to me for thermals), a super-black interior, excellent optical design and execution (minus the mirror gluing), superb guaranteed and certified mirror, and, for me, that cool glossy white tube and mystical Parks logo. The cell was flexing the mirror – I now very strongly suspect the cell was under tension from the collimation locks when the mirror was glued, as outlined above.

Kerry makes a very good point that we should all heed. Even premium 'scopes such as the Parks AZ6 can sometimes come with a few glitches, but once sorted out, certainly produce the goods.

Portability Plus: The Ultralight 8

Vince Rizzo, founder of Infinity Scopes, Indianapolis, produces two exceptionally high quality 'scopes for those who hold portability at a premium. Known as the Uti6 and Uti8, these 'scopes promise to deliver great views in an ultralight package. Here, we'll take a look at the Uti8. This 'scope is a very high quality, superlight 8-inch f/4.5 and weighs 14.7 pounds, including its custom mount and tripod and which, when disassembled, fits into an airplane overhead bin. There is really nothing else quite like it on the market today. It features an ultra-high quality conical primary mirror (we'll be saying much more about these in a later chapter) made by Bob Royce, and a quality matched secondary by ProtoStar. Computer-machined aircraft aluminum and carbon fiber trusses renders the Uti 8 astonishingly light – but certainly not flimsy – and very strong. The mount has built-in sorbothane vibration dampers attached to the tips of the three legs, which do an excellent job reducing vibrations to an absolute minimum.

Comet discoverer David H. Levy is said to have "fallen in love" with the 'scope at first sight, played with it for hours, and concluded you could also pack some clothing in the upper and lower tube assemblies for astronomy trips. Andre Hassid from San Francisco, California, has owned the Uti 8 for a few years now and still raves about its quality. "To my mind," he says, "it ranks in the same league as the 7-inch Teleport (I owned the first one ever made), perhaps even a bit better. It's a very easy 'scope to use and store. The optics are great. I can't tell you if they are equal to the Zambuto or not, as they are very close. Perhaps a more discerning user would find the Zambuto better, but I can't tell the difference. I've owned many telescopes over the years, but in terms of quality, the Uti is equal to the best of them and is very portable. Vince is very responsive and delivers 'scopes very fast."

At f/4.5, the Uti will work quite well without a coma corrector. Eyepieces have to be kept light, though. Hand grenade eyepieces such as the big Naglers will swing this 'scope hopelessly out of balance. Indeed, some folk have eyepiece collections that weigh more than this 'scope. A light hand is also required to use it and not induce excessive vibration, but that is doable with practice, and dampening times are significantly under a second, which is rather amazing. As you'd expect, it makes a mighty fine travel 'scope. You can buy it with setting circles and a battery-powered LED homing mechanism that works off an infrared-connected Palm Pilot with several databases built in that can be added to and expanded. As you'd expect, this kind of telescope technology doesn't come cheap ($3,825 for the Uti8 and $2,495 for the Uti6), but if you want large aperture in an airline portable package, then one of these 'scopes may just fit the bill (Fig. 7.3)!

Another option for the extreme portability enthusiast is the Teleport7 Dob (Fig. 7.4).

This has got to be one the most ingenious telescopes ever designed, and it's a testament to the creativity and engineering talent of Tom Noe, based in Wylie, Texas. The Teleport 7-inch f/5.6 telescope is best described as a folding Dobsonian. As is implied by the name, the 'scope is designed to fold into an airline carry-on size. Collapsed, it stands just knee high. The instrument easily expands to full size on firm tripod struts supplied by Bogen. It uses a Zambuto primary, a 1.3-inch secondary with enhanced aluminum coatings. Add to that a 1.25-inch helical focuser, an integrated Lycra spandex shroud, a secondary heater, and an integrated eyepiece case, and you can see why so many Dob lovers are attracted to the Teleport 'scopes.

Fig. 7.3. Infinity's'Uti2 8 ultralight 8-inch 'scope (Image credit: Vince Rizzo).

The whole idea behind this 'scope was to make it airline compatible, and it meets that goal – but just barely! Weighing 20 pounds it is a bit heavy to lug around. Some owners have reported that it takes a bit of effort to get the truss poles of the 'scope to extend, but once you get the hang of it, it is pretty pain free. Owners report that they can sometimes become a bit stiff or worst still, catch, making extension of the poles that little bit more difficult.

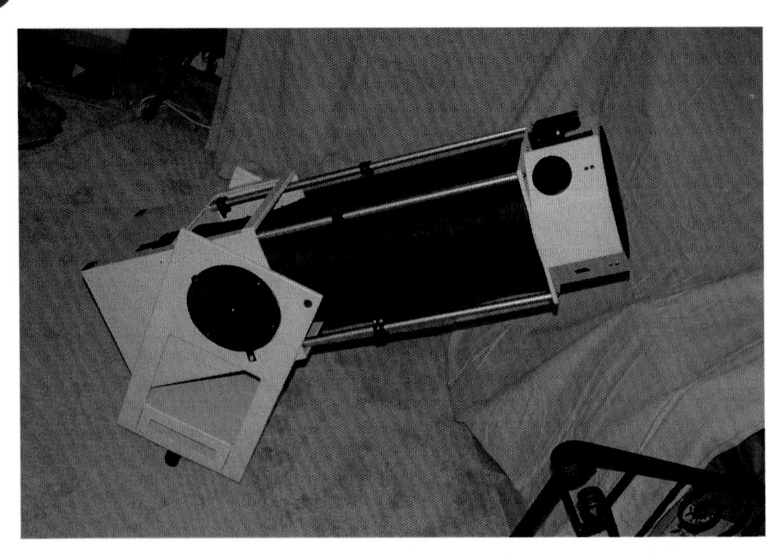

Fig. 7.4. The superlative 10″ Teleport Dobsonian. (Image credit: Bruce Prager).

Care must also be taken while opening and closing the 'scope not to let the built-in light shroud get in the way. The 'scope is just high enough off the ground when unfolded to allow comfortable viewing with a typical observing chair. The included Rigel Quickfinder is a really nice addition and even has its own little place for storage at the base of the unit. Although you can pick up dew heaters for a small additional cost, it is very reassuring to see that the Teleport actually has a secondary heater built in. As all experienced observers know and understand, one of the first things that goes wrong when looking through any Newtonian is that the secondary mirror fogs up, ruining your observing program. How nice it is that the designer of the Teleport 7 has already thought the problem through for you! It's probably overkill if you live in a low-humidity climate, but in other places, it would be an absolute necessity. You can also purchase an optional folding "tri-stand" that elevates the 'scope a further 6 inches off the ground if you're worried about ground thermals or grass dew. All the Teleport 'scopes (yes, there are other models in the 10, 12.5, and 14.5-inch aperture class) can also be retro-fitted with encoders for hassle free push to observing.

Many owners of this telescope attest to the great optics at its heart. The figure on the mirror is well above average. Luna is crisp at 300× in good seeing. Star tests reports all attest to very well corrected optics capable of producing planetary and deep sky views. One enthusiastic owner described how he resolved M13 all the way to the core even with a gibbous Moon hanging in the sky nearby. Re-balancing these 'scopes is a made a breeze by simply adding or removing screw-in counterweights. Consensus is that they are a little overkill, as you can also adjust the tension of the altitude and azimuth movement of the 'scope. Whichever way you look at it, the Teleports are class acts. But you've guessed it, this kind of optical and engineering excellence doesn't come cheap. The current price is $4,500 plus shipping and handling. Even so, the Teleports will appeal to discerning observers who appreciate state-of-the-art engineering in an ultra-portable package. Choose the one that sates your aperture fever and your lifestyle.

A new company, Sumerian Optics, run by Michael Kalshoven, from Vijfhuizen, The Netherlands, offers a range of large aperture Dobs in an ultra portable format. Currently Sumerian Optics sell three distinct types of Travel Dobs; the Propus classic, the Alkaid Compact and the Canopus range.

The "Propus Classic" is a redesigned version of the original Dobsonian Travel Telecope designed by Pierre Stock. The ultra light Propus Dobs can be ordered with either a 10 or 12-inch mirror. The "Alkaid Compact" is a more highly developed, and more compact model of the "Propus Classic". These can be acquired in a 6″ to 14″ format. Finally, the "Canopus" range has a greater emphasis on stability, strength and ease of use. These are offered in 10″ to 20″ apertures.

The prices of these scopes reflect the quality of the optics housed in the lower tube assembly. The basic instruments harbour GSO mirrors, while the deluxe versions can be ordered with mirrors made by Orion Optics UK, with either 1/6 wave or 1/8 wave p-t-v. Currently, Sumerian Optics only exports scopes to EU countries. On the whole, most customers appear to be very happy with their telescopes.

Works of Art

The Dobsonian revolution means different things to different people. For most, it has provided an opportunity to combine generous aperture with cost-effective mounting materials, resulting in the creation of a cost-effective but highly functional instrument that sates the aspirations of the majority of amateur astronomers. But, there are some dedicated observers who don't just want good optics in a mass-produced aluminum tube. To some amateurs, a telescope ought to be as beautiful to look at as it is to look through. These have brought beautifully crafted instruments to the attention of the amateur community. This section explores some of the instruments that fall into this "category" and the motivations behind their creation. Of course, by its very nature, this is a rather subjective topic, as beauty lies in the eye of the beholder.

To begin with, let's pose an important question. Given the ubiquity of good, inexpensive gear, why go to the trouble of designing and marketing hand-crafted instruments? Canadian telescope maker Steve Dodson, award-winning telescope maker and founder of Star Gazer Steve telescopes, has this take on this niche market.

"In the 1990s, he says, "I was able to build, in Ontario, telescopes that were half to one third of the cost of competing 'scopes providing similar observing experiences. The combination of price and performance created strong demand. For instance, in 1997, my 4.25-inch f/10 Deluxe Reflector Kit was providing refractor-like images for $229, a fraction of the cost of refractors giving the same performance, and it was the lowest-priced Dobsonian on the market. A global "sea change" affecting all kinds of manufacturing occurred late in that decade. For instance, it became apparent that to any follower of the news media that the market for running shoes was almost exclusively dominated by goods manufactured by the low-wage economies in Asia. At the same time textile and clothing manufacturers in the West were closing down at an alarming rate. Until 1999, Oriental telescope imports were limited to cheap, barely functioning but "glitzy" 'scopes from China, and excellent but very expensive telescopes from Japan.

"In that year, fairly good Dobsonian 'scopes from China started to pop up almost everywhere. An import boom fueled by the $2-a-day work force overseas was underway that completely re-shaped the telescope market. Traditional and new North American companies imported these 'scopes and put them out under their own names (Celestron, Meade, and Orion, for instance). I think it is safe to say that all of the companies that can ship off-the-shelf Dobsonian 'scopes at low prices make heavy use of Asian (non-Japanese) manufacturing, importing either the complete 'scope or major sub-assemblies. It is no longer possible for a North American enterprise not making extensive use of imports from Asia to offer instant delivery of the lowest cost 'scopes."

You'd think the onset of globalization to the Dob market would ring the death knell for companies such as Steve's. "My market did not disappear," he informed me. "It became a specialty market for discriminating clients who prefer hand-crafting to mass-marketing of assembly-line instruments. Increasingly, these clients found out about my telescopes when they were referred to my website by books and astronomy websites. These clients want a telescope that offers a better experience of sky observing. Offering a better experience is a successful strategy when others have a manufacturing cost advantage. Many of these clients also appreciate having an instrument designed and built by an amateur astronomer with decades of observing experience using materials chosen for performance, rather than by a marketing department and mass manufacturing using the lowest-cost materials."

These are powerful words from a man who has learned to weather the market storm. So what kinds of luxury products does Star Gazer Steve offer?

Lightweight Elegance

His 6-inch f/5 truss tube is a perfect example. With its well figured parabolic mirror, Steve offers above average performance in a small package. The instrument, which can be assembled with simple tools in just a few minutes, combines extreme portability with the classic performance of a 6-inch 'scope, elegance, and exceptional structural and dynamic stability (Fig. 7.5).

Fig. 7.5. Star Gazer Steve's elegant 6-inch f/5 truss tube Dob (Image credit: Patrick Dodson).

This 6-inch Dob weighs in at only 29 pounds when fully assembled, yet possesses the rock-solid feel reminiscent of large truss-tube 'scopes (discussed later in the book) and moves with exceptional smoothness on its Teflon bearings. As we saw in Chap. 2, truss-tube designs allow the quickest possible cooling of the optics and "tube air" to the ambient evening temperature, which provides steady detailed images to emerge more quickly. Indeed, some solid-tube 'scopes take 2 h or more to cool down before getting those terrific views. Unfortunately, such an open tube design means that stray light, dust, and grime can take their toll on the telescope's optics That's when a tube shroud becomes a real bonus.

Such a short, optically fast Dobsonian would put the eyepiece at much too low a level for comfort for the majority of people. But that's where Dodson's genius for innovation shines through. "I created a special, highly rigid tripod, featuring wide triangles, giving a tetrahedral shape, one of the most rigid of structures," he says, "placing the eyepiece within a very convenient height range (46–56 inches)." The focuser on this 'scope is also unique to the genre, consisting of a 2-inch machined aluminum "virtual helix" with a 1 1/4-inch adapter. Instead of turning a focus knob, you rotate a silky smooth bearing.

And, as you can see, when coupled to its "tripod" it is very fetching to the eye. Star Gazer Steve's 6-inch f/5 truss tube is more expensive than those produced by larger companies, but at $887 plus shipping, you still get that reassurance of quality that only a hand-figured instrument can provide (Fig. 7.6).

The company also produces a similarly designed 10-inch f/5 truss tube 'scope ($1,299 plus shipping) for those who prefer more resolution and light-gathering power.

If you like the look and feel of Star Gazer Steve's wooden truss tube Dobs, then you're going to love the 'scopes fashioned by fellow Canadian Normand Fullum. Based in Hudson, Quebec, Fullum designs instruments of outstanding physical beauty. Using only fine woods for the optical tube and telescope mount, each telescope is finely crafted to provide a pleasing eyeful, whether you are looking at the telescope or through the eyepiece. Precision artisan craftsmanship is present throughout the entire construction of the telescope, from the grinding and hand figuring of the primary mirror through to the layers of finish applied to the wood. Indeed, it's rather like a fine piece of furniture built round a Newtonian reflector (Fig. 7.7)!

You'd be mistaken for thinking these 'scopes are all about appearances. But Fullum is no stranger to the optics workshop. This guy can grind mirrors like the best of them. Indeed, one of his finished mirrors was advertised on Astromart with a 36-inch aperture featuring a 1/9 optical

Fig. 7.6. Star Gazer Steve's 6-inch f/5 truss tube atop its elegant tripod (Image credit: Patrick Dodson).

figure and 97 percent reflectivity. Nowadays, Fullum is busy working on some seriously large mirrors for a new breed of monster Dobs soon to hit the market. But if you ask nicely, perhaps he'd still build you the 'scope of your dreams (Figs. 7.8 and 7.9)!

Other amateurs have elected to restore old telescopes and mount them on simple alt-az mounts. A case in point is Robert Katz from London,

Fig. 7.7. (Image credit: Normand Fullum).

Fig. 7.8. Normand Fullum behind his 36-inch f/4.2 mirror (Image credit: Normand Fullum).

Fig. 7.9. A luxury Dob by Normand Fullum (Image credit: Normand Fullum).

England. Robert has a wonderful 10-inch f/8 Calver "Dob" that he presses into service from his back garden when conditions allow (Fig. 7.10).

"My 10-inch f/8 Calver reflector looks like an unwieldy beast," he says, "and by any modern standards it is overwhelmingly long. The original

Fig. 7.10. A refurbished Calver reflector on an alt-az mount (Image credit: Robert Katz).

wooden stand had rotted and was missing its slow motion controls when I found it, but luckily Len Clucas, the former professional telescope-maker for Grubb Parsons in Newcastle, had inherited an identical stand and cradle from the late master mirror maker David Sinden, which he refurbished for me. A stepladder is essential for objects over 30 degrees high, and viewing near the zenith is positively dangerous. And yet – climbing up to the eyepiece apart – it is remarkably easy to use. The eyepiece is always in a convenient position, assuming you can reach it; the azimuth and altitude controls are smooth and make tracking easy even at powers of 300×, and the ingenious system of a clamped tangent arm makes rewinding the azimuth screw simple without losing position. Even though it weighs a ton the telescope is also beautifully balanced; unclamped from the slow motions. With a 40-mm eyepiece in the barrel, I imagine it is the closest you can get to the laid-back star-hopping Dobsonian experience with Victorian equipment.

The optics are fine, and because the focal length is actually less than that of a standard SCT, views of deep sky objects are impressive with a low power eyepiece. It comes into its own with the planets, though, and the exceptional opposition night of Jupiter in September 2010 was memorable in many ways. Thanks to good seeing in Southwest London – the telescope is in Hampton Hill – I spent most of the night watching Jupiter turn in exquisite detail using a fine telescope made in 1882 by one of the two great telescope makers of his day, but a telescope so simple that a child can learn to operate it confidently in 5 min.

The "Dob Buster"

You've probably noticed by now that most innovators in the Dobsonian industry were at one time amateur telescope makers. David Kriege, Normand Fullum, and Peter Smitka all had their own ideas about how best to deliver maximum viewing comfort from an alt-az mounted Newtonian telescope. Norm Butler is no exception. An engineer by profession, Norm has been an active amateur astronomer for over five decades and got his amateur telescope making (ATM) bug growing up in Topeka, Kansas, back in the 1950s. After spending time in Hawaii and China, Norm recently retired to San Diego and now indulges his childhood passion to build extraordinary telescopes. To see what he does, let's take a closer look at an entirely hand-built 10-inch telescope he calls the "Dob Buster (Figs. 7.11 and 7.12)."

Norm is an experienced observer who places comfort at the eyepiece at a premium. Being a 6-footer, he didn't want to spend the whole evening crouched over the eyepiece in a stooped over position, especially when looking at objects high overhead. This is sadly the case with a typical 10-inch f/4 or f/5 Dobsonian. So he set out to design a user friendly Newtonian telescope that had a comfortable eyepiece observing height, was lightweight, and could be easily transported and quickly set up for observing.

Fig. 7.11. The 10-inch f/4 Dob Buster (Image credit: Norm Butler).

Fig. 7.12. The single vane securing the secondary mirror (Image credit: Norm Butler).

Fig. 7.13. A close up of the beautifully designed focuser (Image credit: Norm Butler).

The entire telescope is hand-made (it took him 18 months to complete the task) using ¼-inch and ½-inch plywood, glue, and some wooden screws. The same goes for the alt-azimuth mounting, made out of ½-inch plywood and a 12-inch diameter swivel bearing and three Teflon pads. Norm designed a most unusual 2-inch horizontal single stalk "lead screw" focuser for the telescope and three front-mounted brass collimation knobs (on top of each corner section with the fourth brass knob to lock the primary mirror in place) system using four O-rings and a system of wooden pulleys (total eight) located inside the box (Figs. 7.13 and 7.14).

Norm ground and polished the 10-inch f/4 primary mirror himself in the late 1990s, and it is mounted in the top section of the box and about 4 inches above the declination axis. The plywood optical tube assembly can be easily removed from the top of the "box" by removing some securing nuts through the optical tube vent windows on each bottom side of the tube assembly if access to the primary mirror is needed. Here's the really cool bit: the entire 'scope is counter-weighted with a 16-pound bowling ball, which offers exceptional stability during use (Fig. 7.15).

The 10-inch f/4 "Dob-Buster" is an exceptionally user friendly Newtonian telescope. Consider the finder, for instance. It's an offset,

Fig. 7.14. Check out the 30-mm "periscope" finder affixed with magnets just above the main focuser (Image credit: Norm Butler).

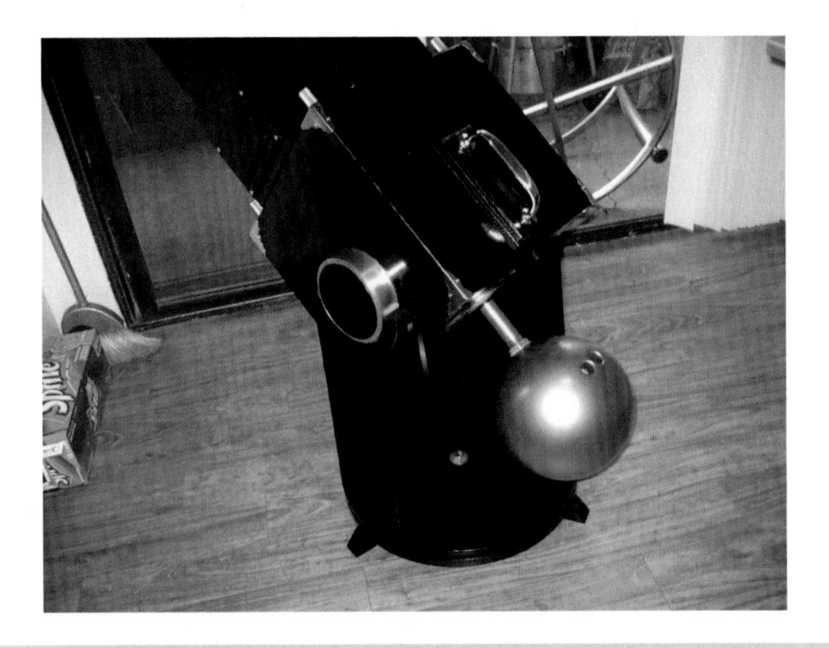

Fig. 7.15. The ingenious "bowling ball" counterweight (Image credit: Norm Butler).

periscope-like 30-mm finder attached to the "Dob Buster" just above the horizontal focuser using ultra-strong neodymium magnets.

What's more, it's just a few inches away from that of the main telescope. Have you ever stopped to consider just how inconveniencing it is to have to constantly reposition yourself between the main telescope and the finder during your observing programs? The eyepiece observing height at the Zenith is a comfortable 5 ft. "I like to chase asteroids and do some comet hunting, says Butler. "My 10-inch f/4 Dob-Buster has a wide field of view and enough light-gathering power to find them (at least some of the ones I'm interested in that are around 10th to 14th magnitude). Once I locate their known position, I can then watch their movement verses the background stars. In terms of deep sky observing, I find many faint fuzzies and nebulae under dark skies with the Dob-Buster. The fainter the better. Maybe one of these days I might get lucky and sweep up a new comet with it. Then I may change its name to "Comet Buster." Now wouldn't that be great! I don't use a coma corrector. I can tolerate some coma with my f/4, especially after observing all night with a good cup of fresh coffee nearby. At some recent star parties, I got a kick seeing the expressions on the faces of some of those Dob owners who watch me set up my Dob-Buster. They don't know quite what to make of it, until they see it in operation!

When a Passion Becomes a Business

It is clear that Norm Butler has created an exceptionally beautiful, hand-crafted instrument that turns heads. He hasn't patented the design, nor was his motivation for building it based on any commercial gain (not that it has no commercial merit). But, some amateur telescope makers have gone one step further and launched a business in this niche market.

How difficult is it to sell the "luxurious Dob" idea to a skeptical and cash-strapped consumer economy?

John Compton from Pena Blanca, New Mexico, has done just that. John was asked about his motivations for designing such luxurious instruments, given the ubiquity of mass-market models.

"I didn't really set out to design a luxury telescope," he says. "At a start over point in my life, an old friend who knew of my longstanding interest in such matters suggested that I make telescopes. The idea resonated with me, and although I didn't know it yet, in many ways I already knew what my telescope would be like, in that I knew what its most important

Fig. 7.16. The New Classic 8-inch made by WoodenTelescopes.com (Image credit: John Compton).

and most consequential parameters should be. Following them to their interconnected conclusion required a lot of test builds to find out for sure what worked, what required untenable compromises somewhere else in the system, and what didn't work. In the end, I learned how very much I've been thinking about telescopes since I got my first one (a mail-order $30 Edmunds 3-inch reflector, in 1963 – at 13 years old). There clearly is a great, possibly universal, deep seated human curiosity about the sky, and everyone wants to look at it through a nice telescope. I've never met anybody who didn't. I used to fly hot-air balloons for a living and was always amazed to find so many who were indifferent to it. Not so with the telescope. Everyone wants to look (Fig. 7.16)."

New Classic 8

What did John have in mind when designing his wooden Dobs? "The overarching goal," he said, "the prime directive, as it were, was to make a telescope that everybody and anybody could use. I didn't want buttons or batteries. I didn't want exacting collimation and leveling procedures. I didn't want electric motors or extension cords, I didn't want computers and I didn't want to find, or even need to know how to find, set-up stars. I wanted excellent optics and a high quality, easy to use finder system. I wanted a rotatable and balanceable solid tube held in a smooth, sturdy mount. I wanted the 'scopes to be self explanatory, intuitive, and

sturdy enough to be used for generations. But most of all, I wanted to make sure they were used by making them more usable.

In short, I wanted a 'scope that naturally delivered that "WOW" sensation people get when they get their first good look at Jupiter's moons and Saturn's rings.

This is, of course, the issue that Dobson addressed with his original 'scopes. He extolled them, quite properly, as something enabling anyone to see the universe. Maybe the most important part of his simple, elegant design and use of materials was to make a lie of the old telescope-building saw (it was in all the 'scope books of the day) that all alt-az mountings were inferior.

Today's commercial Dobs have naturally evolved from the original. Simplified and fancy Dobs abound in the marketplace and do deliver good value. There is a fine line, however, between "inexpensive" and "cheap" that manufacturers naturally try to stay just to the inexpensive side of. I didn't want to be making design decisions based on the drawing of that line, so I was determined to look at Dobson's design in terms of simple functionality, rather than as a design that could be made cheaply. I wanted to see it in terms of the ease of use, that design simplicity delivers. In short, I looked at Dobson's design as simple elegance rather than as a cheap way to do things. Without detailing the arguments, my "ease of use" philosophy eventually led to a decision to make the 8-inch f/6 Newtonian reflector on a modified Dobsonian base that you see on my website. This size is as large a telescope you can have without running into escalating optical and logistical problems. I think it's the best all-around size for a personal 'scope.

Design Considerations

John then told about how his previous experience in another niche market helped him to consolidate the design choice for his wooden Dobs. "Years ago, I flew and sold hot-air balloons," he said, "and once had occasion to listen to a talk by Don Piccard, who discussed the size of the lines that connect the balloon basket to the fabric envelope. He made the point that the Federal Aviation Agency aircraft standards for licensing required lines much smaller than the ones he used. The thin lines were completely safe, but they just looked much too thin for any passenger to feel comfortable. Believe me, in that situation, you really do want a big fat rope holding you up. What I took from that is the importance of a tool "looking like" it will do the job in terms of its interaction with a human user. It's important that a telescope looks like a telescope, especially in terms of the iconic stature telescopes enjoyed across many cultures. A telescope in the background of a TV or movie set says paragraphs about the owner even if it's never mentioned. People know by looking that a telescope ought to inspire curiosity, intelligence, artistic appreciation, and, in general, all good things about human nature, both poetically and scientifically. I always approached this project with a "form follows function" mindset, so I never did anything specific to make my 'scope look like a 'scope. I did spend a lot of time looking at old and new telescopes in books as well as firsthand, in observatories, hoping that their "scopiness" would imprint on my designing brain.

Changing dimensions on my highly integrated design wasn't easy. I was on my third fully functional test 'scope when I decided that the tube clamp discs were so thin at their fulcrum they looked like they might crack under stress. That one small, less than 2-inch, change slowly snowballed to having to change the dimensions of almost everything else. It took a lot longer than I expected, but I think the 'scope now projects the graceful strength I want. Other influences on the final design came to me from things as diverse as my motorcycle racing days, bookstore management, and the old Spanish missions that I've been studying as a hobby and visiting for years both here in the southwest and down in old Mexico for their buttressed building walls.

The three holes in the circular sides of the clamp mount do exactly what holes in the rear wheel sprockets on a racing bike do. They maintain strength and lower the weight. I haven't yet posted pictures of this, the original style, on the website, but will do so soon. More importantly, thinking about motorcycle racing led to an appreciation of how the cost of getting more horsepower increases rapidly for each new bit of power. A better exhaust pipe can give you several (let's say 5) more horsepower distributed better throughout the RPM range, but the only way to get the next 5 HP (if indeed it is possible) would involve extensive, and expensive, engine modifications. That's the law of diminishing returns. That said, even the smallest impediment will influence matters. For example, book sales are very much affected by convenience, so that even shelf placement makes a big difference. People don't want to aim their head let alone stoop. With telescopes, I'm convinced that having to perform even the simplest step, such as finding north or aligning the optics, prevents many newcomers from getting to look at the sky.

In the beginning, one of the first things I did in my shop was to build a 12-foot-long bench for my table saw. When done, I remember standing in front of it and asking myself,"OK, John, what do you want this thing to look like?" Surprisingly, perhaps due to my immersion into all things

telescope, I got an immediate answer that I wrote down on top of the bench. It's still there. I'm not exactly sure what it means, but the look I wanted was, "A Rocket Train to Mars." One of the comments made when I showed earlier this year at an art fair in Austin, Texas, was, "Too cool.... very steam punk," which, I think, very nearly means the same thing.

It stands to reason that the shop I make the 'scopes in must influence the final product at some level, so I want to mention a couple of things about it. It's a long, narrow roofed porch open on one narrow side to the outdoors. There are walls down one long side and across the skinny rear. This leaves one very long side that I fill using fabric from my old hot-air balloon. On cool or wet days I let it hang down as a cloth wall, while in hot weather (I live in the desert) I prop it up and out to give me a nice cool shaded area in which to work. I find the red, gold, and purple light that shines through to be most pleasing, although I am forced to go to the sunshine when judging finishing colors. Most of the sanding is done there or under the pleasant shade of a large old apricot tree. Things that must be done inside are done in a room of the old adobe house I'm refinishing.

Summing things up, I wanted the 'scope to be easy and intuitive to use. To this end I wanted it to be lightweight, maneuverable, have good finders, be human scaled, have professional quality, diffraction limited optics, a solid mount, eyepieces with enough eye relief to accommodate glasses and made of long lasting, quality materials requiring no batteries. In the end, the luxury and beauty of my new 8-inch 'scope flowed, in roundabout ways but inexorably, from those original parameters. I did fail my original desire to not use batteries. I probably experimented with more different finder systems than anything else and in the end liked the ease of a green laser (in spite of the battery it requires). A good 8× or 9× finder 'scope that could do the finding even if the laser battery failed is also included.

The tube is a good example of form following function. I wanted it to be made of wood because it's such a physically and aesthetically perfect material for making something that reaches from Earth to the sky. It had to be cylindrical, to enable rotation in its clamp, which led me to a long period of tube making. The best ones used boat-building techniques and really looked good but just took too long to build. I finally settled on my Dodecatube™ slat-and-rib design that not only didn't require complicated curves but also nicely echoes the twenty-four 15-degree time zones on Earth as well as the twelve signs of the zodiac.

The dimensions of parts were largely dictated by the design, but when possible I always tried to give some extra meaning to the final shape and dimensions of things. There was some leeway in the placing of the eyepiece holders, for example, so I tried to make their height correspond with the width between the buttresses using the ratio of Fibonacci's number. Everything I read about such things seems to either make exaggerated, deep meaning, resonance, and new-agey claims, or knee jerk dismissals of the whole idea. All I can say is that it gave me something to work with when there was no compelling engineering reason dictating doing otherwise. The end result does seem to have an organic look, especially if you squint at its silhouette at dusk. Some people see a dragon/dinosaur.

A Eureka moment for me was getting the clamp/cradle to work. It took longer than anything else. I must have tried several dozen ways to do it but was stumped as to finding an elegant solution. Then, one day while working on something completely different (the jig I use to glue the tube parts together) a slightly undersized washer stuck when I tried to take it off its bolt. It's a common experience that is fixed by simply lining the washer up 90 degree from the bolt, but I immediately knew, after months of trial, that I had hit on the solution to the clamp problem. I well remember the deep pleasure of thinking, "So this is what an Eureka moment feels like." It took several more months to work everything out, but my 'scope cradle/clamp still works on the "washer on bolt" principle. One last manufacturing principal that I found both curious and useful for woodworking was to "measure in inches to the nearest (half) millimeter."

The only promoting I've done is the web page and seven art shows over the last year. I've never considered myself much of a salesman, preferring to concentrate on making a unique, quality product, trusting that the buyers will eventually find me. Truth is, that hasn't happened yet. I haven't sold a telescope, but I remain absolutely convinced that I will. I just have to find a way to get it in front of the right people. As the final form of the 'scope came together and I saw just how good it looked, I realized that it was very worthy of the iconic nature of the thing. That and the reaction of others made me realize that it was art, and I quickly learned from them that selling art has different rules. For one thing, my asking price (based on what a production run would cost per unit) was way too low. A higher price, in effect, gives an art piece more value. Pricing art, it seems, involves making the price high enough to indicate the value of the piece. If an item doesn't cost as much as people think it should, they will think it isn't worth much.

My first showing was at Los Alamos, New Mexico, which I thought would be a natural market, as it's full of scientists working at the national laboratory famous (infamous) for the development of the atom bomb. I was immediately surprised there at the stereotypically different ways artists and scientists react to my telescope. The artists love the way it looks and are amazed that I was asking so little for it. The scientists also like the way it looks and appreciated the functionality and optical quality and but often seem to think the extra cost over cardboard tubes is largely wasted.

They wanted bigger mirrors, clock drives, and all the extras I eliminated to make the 'scope user friendly. Since I wasn't selling any telescopes anyway I soon decided that I should at least not sell them for a higher price. When asked, I still semi-seriously tell people that I have two prices: $3,800 for scientists, $5,800 for artists.

Actually, not one person has ever flinched at any price I've quoted (up to $10,000). The almost universal reaction to a price quote is an incredulous, "Is that all?"

The high prices recommended by artists, I realized, reflected their assessment of what a one-off custom 'scope would go for. Since then I've realized that the original concept of making the 'scope easy to manufacture was very much part of the design process, and it was poetically wrong to forget that so I've gotten back to the original intent to sell custom-looking telescopes as an in stock, ready for delivery item, rather than an expensive, custom-made trinkets for the well-to-do. In the meantime, I've been approached by several art galleries wanting to show the 'scopes I'm presently working on. Finally, I do believe that the economic recession has kept a lot of people out of the market. Art sales are generally down all over, and in spite of my opinion, most people continue to consider telescopes as a luxury, and not as a necessity (Fig. 7.17).

Fig. 7.17. It's amazing how the rich colors of Baltic birch play with the daylight (Image credit: John Compton).

John Compton's sentiments will probably resonate with a lot of amateur telescope makers and established manufacturers alike. All we can do, on behalf of everyone who has a passion for telescope making, is to express gratitude to these folk who paddle on, against the odds, to bring their wondrous creations into being. In the next chapter, we'll take a look at the leviathans of the Dobsonian world, telescopes that transcend the normal and mundane and carry the observer into new realms of observing experience. Here be monsters!

CHAPTER EIGHT

Here Be Monsters

What drives me in this venture is the extraordinary splendor of the universe one can experience at the eyepiece. Objects like the Orion Nebula, the Whirlpool Galaxy, or the Veil Nebula are beautiful beyond description in large telescopes. Photographs simply do not do justice to the visual thrill at the telescope eyepiece (although they may show a lot more details than you could ever see visually). The best description of my favorite hobby would be "tourism of the deep sky."

Robert Houdart

Words like these, penned by Belgian amateur Robert Houdart – who regularly looks through a 42-inch Dob – can never do justice to the unbridled beauty of the deep sky and the endless treasures it presents to the curious mind. Once an amateur astronomer gets a look at Omega Centauri or the Orion Nebula (M42) through a 16-inch Dobsonian, chances are he or she will be pining for more aperture. You see, there's really no substitute for it. More light means more photons bathing the human retina and with it, the promise of seeing ever greater detail. Unfortunately, few make the transition to larger apertures, but when they do, few ever look back. This chapter is devoted to the monsters of the Dobsonian world and the observers lucky enough to own and use them.

The sky really is the limit with large 'scopes like these. Planetary nebulae, globular clusters, and faint galaxies appear as if out of nowhere! With them one can clearly make out a string of mottled HII regions in the

N. English, *Choosing and Using a Dobsonian Telescope*, Patrick Moore's Practical Astronomy Series 1, DOI 10.1007/978-1-4419-8786-0_8,

Andromeda and Pinwheel Galaxy, a wealth of structure in the tail of Comet 17P/Holmes as it sprinted through Perseus, and vivid color in the Ring and Dumbbell Nebulae (M57 and M27, respectively). Have you ever seen Saturn on a good night through an 18-inch 'scope? 'Tis a sight for sore eyes. Even the outer gas giants are easily studied with this kind of aperture; several moons of both Uranus and Neptune could be made out near opposition. M81 and M82 present themselves in absolutely stunning detail, and nearby NGC 2976 and 3077 are easy pickings. The globular clusters M13 and M 92 are awe inspiring, seen as if cruising past them, in a spaceship of the imagination, at close proximity.

Through the 18-inch the Crab Nebula looks like a proper invertebrate, but to some it resembles a jellyfish more than a crab. Although resolution does seem much more sensitive to the seeing conditions, during the wee small hours of the morning, you can get up to look at the Orion Nebula

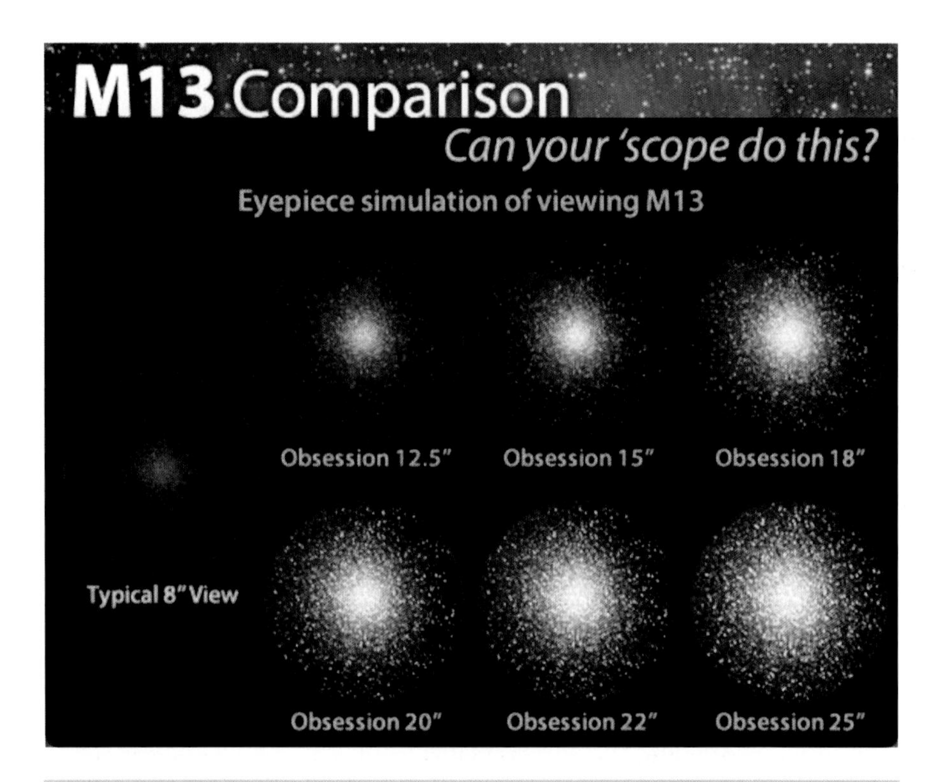

Fig. 8.1. A simulation of what 'scopes of different apertures can "open up" in the globular cluster M13 (Image credit: Obsession Telescopes).

(M42) rising above the eastern horizon. It looks blue-green with swirls here and there of red. Indeed the view is almost like one of those three-dimensional images taken by Hubble. If you want to put the "wow" into visual astronomy, you have to look through one of these behemoths on a good night.

So, what choices are available to the amateur? First of all, these 'scopes don't come cheap. The optics and the tubes that carry them cost bucks, big bucks (Fig. 8.1).

Flying the Flag for Britain

David Lukehurst has been a keen amateur astronomer since he was 10 years old. A past president and Curator of Instruments of the Nottingham Astronomical Society, he started building telescopes for himself back in the 1980s and from there he has slowly evolved into becoming a professional telescope maker. David's inspiration wells from the descriptions of truss-tube Dobsonians he came across while flicking through the pages of *Telescope Making* in the 1980s, especially the designs of David Kriege. Lukehurst now specializes in making his own variations of this type of large, stable, and highly portable truss-tube Dobsonian (Fig. 8.2).

Fig. 8.2. The Lukehurst 20-inch f/4 truss tube Dob (Image credit: David Lukehurst).

Over the years, Lukehurst has gained a solid reputation for delivering optically excellent telescopes in truss tube Dobsonian frames that are solidly built and a joy to use. His current range includes telescopes with apertures from 12 up to 24 inches (although he has built and designed larger instruments). The mountings are made from high-grade exterior plywood and are carefully sealed and painted, making the telescopes durable, attractive, highly functional, and easy to use. One can choose from a Standard or Ultra Portable II range, and you can also specify either a ¼ wave mirror or a 1/10 wave figure on the "Deluxe Model." Lukehurst has nearly all his primary mirrors made locally by Norman Oldham (Oldham Optics), in Scarborough, but occasionally he employs mirrors made by Nichol Optical and Orion Optics UK.

The smaller models have plate glass mirrors, while the larger Lukehurst 'scopes feature mirrors made from low-expansion Schott "Suprax" or Corning "Pyrex" borosilicate glasses or BVC glass/ceramic. The elliptical secondary on the 20-inch f/4 standard model has a minor axis of only 96 mm, giving a central obstruction of just 19 percent.

One enthusiastic owner told of the extensive star testing he had done on his 20-inch f/4 Lukehurst Dob. "I couldn't detect any major defects," he told me. "There was no sign of any astigmatism – a fairly common ailment of cheaply made mirrors – even close to focus. The test revealed near perfect correction and very smooth and high contrast optics. That's especially impressive, since the primary mirror is large and optically fast. My 20-inch can handle very high powers. My first light experience showed me this mirror could easily handle 40 times per inch of aperture on planets. In good seeing conditions Jupiter and Saturn would reveal a high level of detail and remain sharp at magnifications in excess of 600X. Of course, the seeing rarely permits steady images with this size of 'scope, but in moments of calm, amazing atmospheric details jump out at you and reveal the full potential of the quality optics on the instrument."

What's there not to like? Well, of course, this thing isn't light. Weighing 110 pounds, it's sure to require two people to lug it out into the field, but set up from there takes only a few minutes and is still a one-person job. The Ultra Portable 20-inch weighs considerably less, though, and folds away for easy storage and transport to a dark sky site. Some owners have noted that if the truss poles are not sufficiently well tightened it can cause the 'scope to go out of alignment during an observing session.

What is especially nice about this range are the prices. The standard 20-inch f/4 will set you back £3,695 plus shipping; the deluxe version is yours for another grand (£4,650 plus shipping). The Ultraportable II versions of this 'scope are £4,295 and £5,295 for the standard and deluxe models, respectively (Figs. 8.3 and 8.4).

Fig. 8.3. David stands beside his 20-inch Ultra Portable II Dob. (Image credit: David Lukehurst).

Fig. 8.4. The dismantled telescope with altitude bearings folded (Image credit: David Lukehurst).

Orion Optics UK New Mega Dobs

We have already seen some of the great products offered by Orion Optics UK, but recently this tried and trusted English company has entered the monster Dob market by producing its 20-inch OD 500 semi-truss tube Dob. Unlike many other Dobs of this size, the OD 500 is an all metal design. At its heart is a very nicely figured mirror (1/6 wave), with Hilux enhanced aluminum coatings for extra brightness and durability. Weighing in at 60 pounds for the tube assembly and another 25 pounds for the base, that makes it one of the lighter 20-inch mega Dobs on the market (Fig. 8.5).

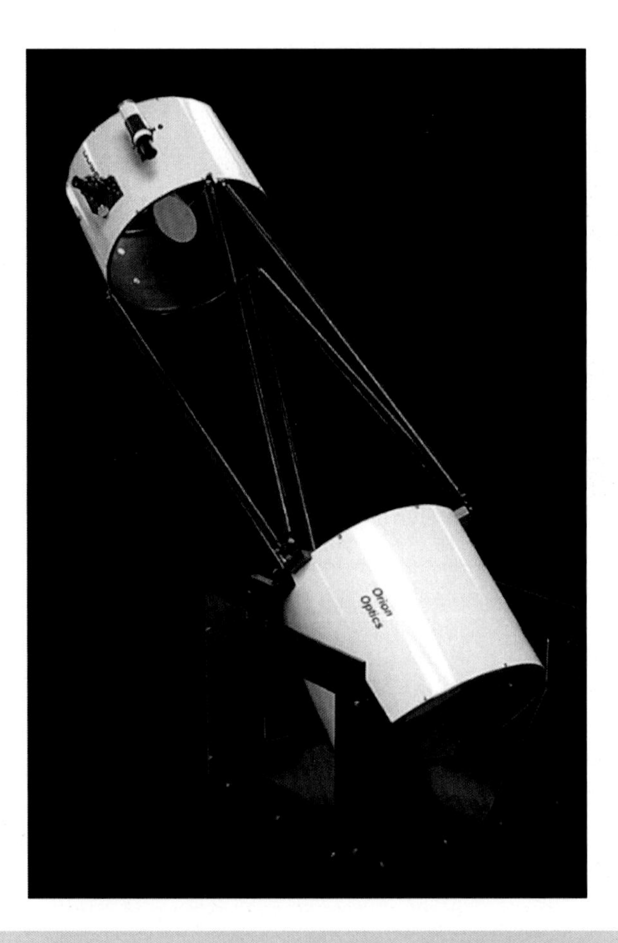

Fig. 8.5. The Orion Optics UK OD 500 semi truss-tube Dob (Image credit: Orion Optics).

So what's it like to look through? One enthusiastic owner, who suffered a severe bout of aperture fever to provide the answer, said: "It is very well engineered and glides effortlessly on its azimuth and altitude bearings even with a large, heavy eyepiece inserted into the focuser. I use my OD 500 'scope for "wide field" scanning most of the time, and as you can imagine, a 'scope this size can pull in a tremendous amount of fainter stars in Milky Way fields. Through a smaller instrument (of course, depending on exact size and magnification being used), some of the Milky Way will be unresolved, but that's not the case for a super-large aperture 'scope like this.

The first thing you have to get used to is the very small field of view, even in a low-power eyepiece. If you use star clusters as an example, sure, you might not be able to fit the double cluster in one field of view, but that's not exactly a design fault is it? In an Orion OD 500, there are literally hundreds of fainter NGC objects that will look just as good as the double cluster does through a 6-inch 'scope. Indeed, to put things in perspective, I have never experienced a better view of it. With my lowest power, wide-angle eyepiece, both clusters just barely fit in the field of view, and the 20-inch 'scope brought in enough light to make all the stars dazzling bright. With really big objects, such as the large Magellanic Cloud, the entire structure takes up several fields of view in this 'scope. So, although it is well nigh impossible to see the entire structure, you have fun panning around and drinking up incredibly fine details – its anatomy, if you like – that are quite beyond the range of regular backyard Dobs. To get the most out of 'scopes this large, it really does pay to get to know the sky intimately first. These are typical results, but it shows why I really don't believe that small 'scopes have any special advantage over big ones (other than the physical advantages, which could be a deciding factor, but that's an entirely personal choice).

Big Dobs from Down Under

Australia has its big Dob makers, too. Peter Read, founder of SDM telescopes, based in Shepparton, north central Victoria, builds custom-made Dobs in apertures ranging from 15 to 32 inches. Deriving his inspiration from David Kriege of Obsession Telescopes, Read makes the tube assemblies from durable marine plywood and stainless steel hardware. All SDM instruments are bespoke Dobsonians of any size and focal ratio tailored to the customers exact requirements. "I only make about ten complete 'scopes a year," he says, "but each one is a handcrafted work of art that is both great to look at and look through. I supply mirrors from the world's most respected opticians, including Mark Suchting, Steve Kennedy, Terry Ostahowski, and Carl Zambuto. My instruments are the only Dobs on the market that come complete with all extras included. Each 'scope is fully assembled and tested under the stars. I can set up the 'scope for either visual only or imaging and visual use.

Every SDM Dob comes complete with carry cases, shroud, storm cover, and transport cover. I am the only Dob builder that has all the wiring and electrics complete and fully self contained. This includes a rear fan, laminar layer fans, auto dew control on the secondary mirror, and built-in heater straps for Telrad, Eyepiece and finder and no loose wires! SDM 'scopes also come equipped with the famous Feathertouch focuser by Starlight Instruments, and we have even devised a unique method of fitting the Moonlight filter slider and offering it as part of the package.

Read also installs encoders and Argo Navis by Gary Kopff on all 'scopes and the Generation 3 ServoCAT GoTo system by Gary Myers. As the owners of 'scopes with wooden clamping blocks have experienced, there are times when the poles become stuck due to changes in temperature and humidity. To counteract that, Read uses Delrin (acetyl) for the pole clamping blocks, which neatly solves this 20-year-old problem.

Fig. 8.6. The SDM 30-inch f/4.8 Obsession-like Dob (Image credit: SDM Telescopes).

One other advantage of SDM 'scopes is their extreme ease of assembly. You simply drop the pole assemblies into the Delrin blocks and tighten. Then drop the upper tube assembly onto the quick action clamps, lock in place, and you're done. Set up can literally take just a minute or two, or you can opt to wheel the 'scope out fully assembled to begin observing. In fact, Read is the only Dob maker that offers split system wheeling handles (Fig. 8.6).

American Behemoths

The United States has produced its fair share of giant telescope builders and justifiably maintain a great influence over the global Dob market today. One of the leading players in the super large Dob market is Eric Webster, founder of Webster Telescopes, based near Detroit, Michigan. Coming from a long line of woodworkers, the Webster family have seamlessly blended fine optics into beautifully designed wooden framed truss tube Dobs. Currently, their largest 'scopes (up to 32 inches in aperture) are outfitted with optics created by Steve Kennedy. The smaller C series Webster 'scopes (from 14.5 to 20 inches) are fitted with Zambuto optics.

All the Webster 'scopes feature nicely finished joinery. Where other manufacturers have assembled their rocker boxes with screws, Webster has replaced them with proper dovetails. The strong but lightweight aluminum trusses and hardware are anodized flat black for greater longevity. Most other 'scopes have painted, foam rubber-coated, or even bare aluminum trusses.

The mirror cell in the Webster 'scope has got to be one of the most innovative parts of the entire instrument! It is both very compact and very low profile, which, in turn, allows for such a small mirror box. What's more, the cell is very well vented for quick cool down times. The main frame of the cell is composed of tubular steel. The welded joints look clean, with a very small amount of splatter evident. The cell is painted flat black. The 32-inch Webster uses a spherical bearing (described below), 27 point cell. Each triangle is made from 1/3-inch thick aluminum with a stainless bearing – a type of Teflon-lined joint that allows free movement in any direction – bolted to each corner, giving enough clearance for air to ventilate properly beneath and across the mirror surface.

A stainless steel stud runs through each triangle and into a spherical bearing that is pressed into a 3/8-inch thick steel bar. This bar, in turn, has a larger spherical bearing placed in its center and engages the primary mirror's collimation knobs. Each and every moving part of the mirror cell floats around on these ultra smooth bearings. The triangles are kept in alignment with a thin stainless steel wire in place of a wide plastic ring seen on competing models. The wire would allow much more effective airflow to occur. And cooling off a big mirror is not as easy as it sounds.

The Websters are fond of gluing down optics to avoid using a sling for lateral mirror support, but some of their 'scopes come with a stainless steel cable sling. Cable slings have been around for a while, but this may be the first time one comes fully factory installed in a commercial telescope. Cable slings have gained popularity with discerning Dob users because they reduce the stress near the face of the mirror and thus do not stretch or shrink after sitting as the temperatures change. The thin cable allows hundreds of times the air exchange rate over the fabric blanket compared with what a conventional sling provides. The sling is adjusted by hand, so no tools (other than a measuring tape) are needed, but realistically, one would hope that this has to be done already at the factory (Fig. 8.7).

It is remarkable how easy a 32-inch 'scope is to assemble. In fact, the giant Webster 'scope can be rolled out of the garage or assembled by one person in just a few minutes! Eyepiece height at the zenith is only 108 inches, so if you're a six footer you'll need just a few steps up to observe objects passing high overhead. The focuser is of a high quality, Feathertouch, two-speed variety. And finally, the price: this superbly engineered 32-inch f/3.6 instrument will set you back the princely sum of $27,099 (plus shipping).

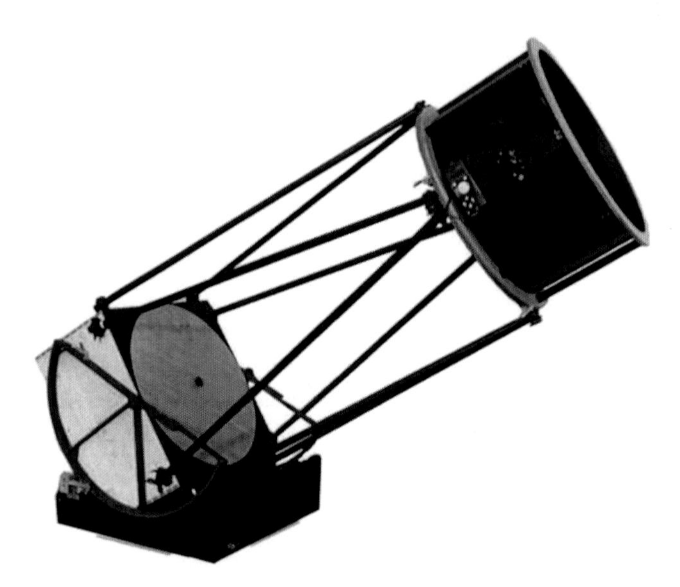

Fig. 8.7. The Webster 32-inch f/3.6 Dob (Image credit: Eric Webster).

The AstroSystems Range

AstroSystems, a small company based in Lasalle, Colorado, also specializes in manufacturing large "primo" Dobs for amateurs suffering from extreme aperture fever. Presently, the company produces instruments ranging in size from 16 to 32 inches in apertures. All the Astrosystems 'scopes are, not surprisingly, truss tube designs to minimize the weight and cost of transport and setup in the field. Their truss tube Telekits have been thoroughly redesigned to meet the new trends in sub f/4 optical systems (Fig. 8.8).

According to AstroSystems, optics that are figured at f/3.3 to F.3.9 require an entirely new approach to telescope design. For instance, consider the secondary mirrors. To properly support the big, heavy secondary mirrors on this new breed of monster Dobs, companies were compelled to introduce several design changes. For instance, their 'scopes have larger holders, with proportionately thicker shells and a ½-inch mounting stud. The adjustment plate has been increased to 3/8-inch thick, with a rigid foam insert used to position the mirror within the holder accurately and so keeping its position regardless of where it is pointed in the sky.

Remarkably, there are no tools needed to adjust mirror holders of 2.5-inch or larger in minor axis size, as they come with thumbscrews as standard. All the secondary mirror sizes feature the company's innovative

Fig. 8.8. The AstroSystems' monster 28-inch f/3.6 Telekit (Image credit: AstroSystems).

four screw/two-axis adjustment that greatly simplifies collimation. Two nuts with nylon washers position the holder laterally and also allow easy adjustment while changing the orientation of the secondary during collimation. The mirror itself is securely held in place against the front lip of the aluminum bezel with polyester batting on sizes up to 4.0 inches and a rigid foam insert on sizes over 4.25 inches. Offset is included, necessary to evenly illuminate the field in short focal length telescopes.

Other nice features include the addition of a Moonlite CR-2 focuser as standard equipment on all the Telekits. If you opt to put one of their giant Dobs together on your own and are handy with woodwork, you can make a primo Dob yourself, cutting the cost to just 50–70 percent of what it would ordinarily cost to order one up completely fitted. You also have a choice of the optics you wish to employ in your AstroSystems 'scope. You can choose from tried and trusted names such as Galaxy, Pegasus, or Ostahowski optics.

Whichever way you look at it, a great deal of time and thought has gone into designing and constructing these "primo" Dobs. The AstroSystems telescopes exude quality, both in terms of the quality optics they employ and the components out of which they are constructed. As one enthusiastic reviewer commented, "ultimately, there isn't a single component on this telescope that doesn't scream quality. This quality does come with a steep price, however. For the cost of a finished telescope with optics, the cost can be as high as three 16-inch LightBridges. This begs the question – is the finished TeleKit three times better than a LightBridge? Ultimately, yes. You get what you pay for, and in this case, the quality justifies the price. This telescope has possibly the finest attention to detail and brilliance of design of any Dob. It gets the highest recommendation!

A New Kid on the Block

At the 2010 North East Astronomy Forum (NEAF), a new Michigan-based company, Great Red Spot Astronomy Products, launched their new monster Dob, a magnificent 40-inch f/3.6 instrument. Needless to say, the 'scope created quite a buzz with attendees. The giant mirror, made by Mike Lockwood, is of unquestionable quality. Unlike other super large Dobs, the designers of the optical tube went back to the drawing board to incorporate a number of safety features on the telescope to prevent damage to both the optics and the sensitive electronics. Company founder Al Murray noticed a disturbing trend with some telescopes that have the primary mirror totally exposed. "Owners of these telescopes quickly find to their horror that people walking nearby kick up small stones out of the grass," he says. "You hear them pinging off the face of the mirror all night long.

We ran the numbers, and adding protection around the mirror increased the weight of the 'scope by a mere 28 pounds – a small price to pay for the security of a $40,000 mirror. The mirror enclosure also blocks stray light, giving higher contrast images at the eyepiece."

The same goes for the secondary mirror. Its back and sides are completely enclosed, giving it maximum protection from dust, dew, and other sources of grime. This design allows you to set it on the ground without damaging the secondary mirror. Most mirror holders are made from plastic, but the Great Red Spot Dob is composed of machined aluminum. What's more, the secondary mirror (with an impressively low 18 percent obstruction) is fully offset to ensure easy collimation and minimal vignetting of the eyepiece field of view.

The company's attention to safety issues doesn't end there. "Another disturbing trend we see is that some manufacturers have been installing the ServoCat brain on the outside of their telescopes, leaving the wiring open to damage," he says. "Even worse, some have the face of the ServoCat sticking out of the telescope body at ground level. One can too easily step on the connectors and break them off flush with the face of the control box! In addition, placing the computer ports at ground level, where they can collect dew and dirt, is an alarming design flaw. We designed the wiring for minimal exposure. Tucked out of harm's way, the ServoCat has no connectors exposed or placed anywhere they may be stepped on."

There has been a tendency in recent years for big Dobs to use carbon fiber as the material of choice in the design of the truss poles and the upper tube assembly. Murray has departed from that tendency, returning instead to more a durable aluminum substrate. The primary mirror cell provides 27 points of support, with the movements completely guided by spherical bearings, ensuring a slop-free system. The altitude bearings – which in this case are fully 52 inches in diameter – are also made from lightweight cast aluminum, which has greater mechanical strength than plywood laminate used in other big Dob designs. As a result, the 'scope can be used without using counterweights, especially when changing from light to heavy eyepieces in the field.

Not to be outplayed, Orion USA has launched its own range of ultra-portable mega Dobs. In the 20-inch class, Orion has introduced the UP20, its 'premium, ultraportable, truss tube Dob. Featuring a diffraction-limited, f/4.2 parabolic mirror, this 'scope has an eyepiece height of 80 inches when pointed at the zenith. Despite its all metal construction, the 'scope when fully assembled weighs in at an impressive 96 pounds, considerably lighter than those provided by other manufacturers. Of this, the mirror is the heaviest component (57 pounds). Instead of the traditional six pole truss design of other models, the UP20 has eight poles for extra

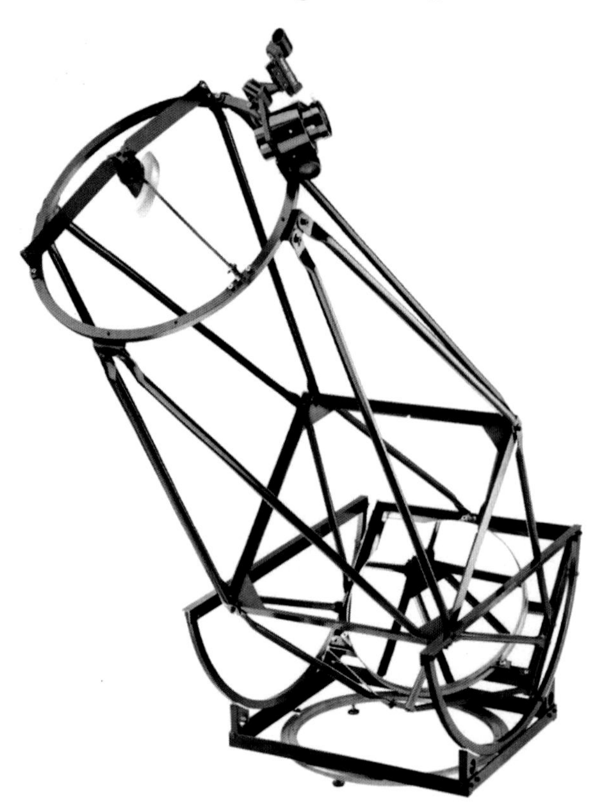

Fig. 8.9. The 20-inch f/4 Orion UP20 ultra-portable premium Dob (Image credit: Orion Telescopes).

rigidity. The whole thing looks like something you could build out of a scaled-up Mechano set (Fig. 8.9).

This 'scope has a number of other nice touches. What is especially noteworthy about this Dob is the actual structure of the primary mirror itself. Orion has adopted a so-called "sandwich mirror" technology that features two solid glass plates separated by a thermally optimized "open core," which consists of pillars of fused glass between the plates. This has the neat effect of allowing the air to circulate freely in between the layers of the mirror, apparently enabling the mirror to reach thermal equilibrium 10 times faster than a traditional solid 20-inch mirror and means that the mirror can deliver good images more rapidly. Even so, you'll definitely need to install cooling fans to accelerate thermal acclimation with this size of 'scope.

What is most striking about Orion's detailed description of this telescope is the extent to which state-of-the-art computer-optimized technology has been used from its inception. Indeed, Orion seems to pride itself in declaring that "these telescopes weren't made in a garage or wood shop." That statement can mean different things to different people. That said, there's certainly a lot of little things about this 'scope that appeal to the mind. It's got an effective baffling system in place to retain maximum contrast, a precision Crayford-style focuser, enhanced aluminum coatings on the mirror, oversized azimuth ball bearings, over 25 inches in diameter, using an "exclusive" laminate altitude-bearing material. But, you've guessed it, all that technology comes with a steep price tag. Expect to pay somewhere in the region of $8,300 for the privilege of owning one!

If you thought the company was ambitious in launching a mechanically super-high tech Dob in the UP20, then wait until you hear about the giants Orion will shortly bring into production. When you scale up the same technology you can get meter-class instruments capable of bringing details beyond the ken of the vast majority of backyard observers into full view. You could, for instance, easily see what the third Earl of Rosse strained to observe through the eyepiece of the Leviathan of Parsonstown.

Admittedly Lord Rosse, who lived at a time when the best mirrors were made of speculum metal, had a 72-inch mirror at his disposal, yet the 36-, 40-, and 50-inch instruments heralded by Orion promise to show you more than the Leviathan of Parsonstown ever could. These monster Dobs have monster price tags to boot; the 36-inch will set you back a cool $55,600, whereas the 50-inch can be had for a heart-stopping $123,000. For that price, you'll get premium mechanics with premium optics supplied by award-winning telescope maker Normand Fullum. Here's some of the design features built into these monster 'scopes:

- Enormous light-gathering power. The 36-inch mirror collects over 1,000 square inches of light – nine times more than that collected by a 12-inch telescope!
- The primary mirror is a high-quality f/4 optical system gives a whopping 144 inches (3,658 mm) focal length to deliver exceptionally sharp, bright images with generous image scale. Fullum assures pontential buyers that the figure will be at least 1/8 wave at the eyepiece.
- Enhanced (96 percent) coatings on the primary and secondary maximize light throughput.
- The primary mirror is made from low-expansion borosilicate glass molded in a honeycomb mirror blank design to make it exceptionally strong, yet light, and assures good optical performance, even as the temperature drops. It is mounted in a 27-point flotation cell and an eight-point wiffle-tree edge support system, which prevents mirror deformation, thereby maximizing image fidelity.

- Transportable but not portable. It is constructed entirely from aluminum and aluminum/Russian birch composite structures. The complete system weighs less than 400 pounds, with the heaviest component being the primary mirror (~ 150 pounds).
- It features complete tracking and GoTo capability, making locating and observing faint fuzzies easier.
- Built-in wheels allow you to roll the telescope in and out of a shed or garage with ease. These wheels can be removed once the desired location for the 'scope has been found.

These gargantuan telescopes will be available from May 2011 (with each requiring a 75 percent down payment).

That completes our survey of the current Dob market. Needless to say, we covered quite a bit of ground on our journey from tiny 3-inch toy 'scopes to monolithic 50-inch instruments towering head and shoulders over the landscape. In Part II of this book, we'll be exploring how to squeeze the very best performance out of your Dob, however humble or magnificent it is.

PART TWO

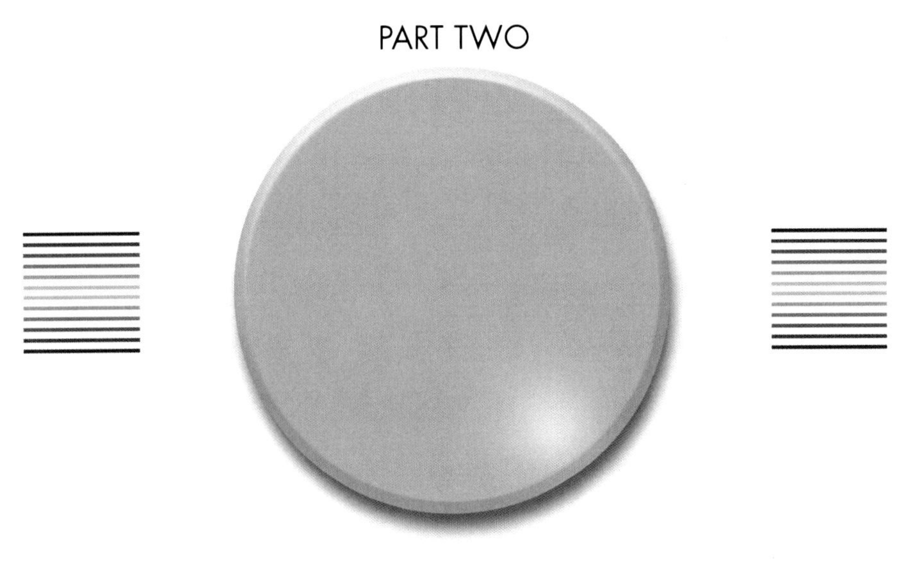

Using Your Dobsonian

CHAPTER NINE

Accessorizing Your Dob

Most Dobsonian telescopes tend to be shallow dish Newtonians and, by implication, require care and attention if they are to perform optimally. Because many of these telescope have such fast focal ratios (from f/3.3 to f/4), Seidel aberrations such as coma, distortion, and field curvature can be problematic, but a good optician ought to be able to reduce spherical aberration and astigmatism to very acceptable levels, even for a fast mirror. In this chapter, we'll take a look at some the key accessories you will find useful while using your Dob in the field. In the end, though, as discussed previously, optical quality starts with the quality of the mirrors in your telescope. Compromise on that, and you risk compromising the performance of your instrument no matter how good the accessories are.

The Curse of Coma

As we saw back in Chap. 2, coma makes stars look like blurry comets whose tails point out radially from the center of the field. The faster your 'scope's F ratio, the worse the effects become. All parabolic mirrors used in Newtonians are limited in field sharpness due to coma. Even a perfectly made f/4.5 parabola has a relatively small, diffraction-limited field. Eyepieces offering low and medium powers in a fast Dob are much more

N. English, *Choosing and Using a Dobsonian Telescope*, Patrick Moore's Practical Astronomy Series 1, DOI 10.1007/978-1-4419-8786-0_9,

affected by coma than are short focal length (high-power) eyepieces giving high power. That's because the latter have very small fields of view over which the coma is rarely severe. Although a big Dob can be charged with very high powers to look at the Moon and the planets, seeing conditions rarely allow them to. It is no mystery why the majority of Dobsonian enthusiasts use the great light-gathering power of their instruments at low and moderate magnifications most of time.

Previous attempts to correct for coma have met with limited success. Some correctors were designed mainly for astrophotography. Another approach was to design dedicated coma correcting eyepieces. However, these were limited to relatively small fields of view (typically 50 degree apparent fields) and so couldn't be used with modern wide angle eyepieces. It was these limitations that spurred Al Nagler, founder of TeleVue Optics, to look into the matter and which eventually led to the introduction of his Paracorr. This is what Nagler had to say on its development:

"I decided to try to achieve the same field quality for parabolas as reached by my Nagler eyepiece and four-element refractor designs. Existing coma correctors seemed to be limited in color correction, spherical aberration, and convenience in use. Paracorr (parabola corrector) uses two multi-coated, high index doublets, is completely color-free, center and edge, and installs like a Barlow. Coma is corrected so well, the diffraction limited field area of an f/4.5 Dob/Newt is increased 36 times (Fig. 9.1)!"

So how does the Paracorr work? In short, by adding coma of the opposite sign to that produced by the mirror. In doing so, it also increases the focal length by about 15 percent, thus having a mild Barlow lens effect. It is a four-element design, in which two elements eliminate coma and the other two reduce any introduced spherical aberration.

How well does it work? Very well, judging by the many thousands of these units now used in the field by fast Dob lovers. The original Paracorr works optimally in f/4.5 systems, reducing the coma to the levels normally encountered in an uncorrected f/8 mirror. That said, there are limitations to its coma-busting powers, especially when 'scopes get faster than f/3.5. For faster 'scopes still (such as those sold by Starmaster), a more powerful coma correcting device is wanting. At the Okie-Tex Star Party of 2009, Nagler brought along his new prototype coma corrector, the long-awaited Paracorr Type 2, to evaluate its performance in one of Rick Singmaster's 22-inch f/3.3 Dobs. Actually, just as the original Paracorr was optimized for f/4.5 optics, so, too, is the Paracorr Type 2 ideally suited to f/3 or faster 'scopes. At the Winter Star Party of 2009, Mike Lockwood tested it out on his 20-inch f/3 "no ladder" Dob, and he was very satisfied with its performance. By all accounts, the results were very encouraging, and amateurs are eagerly awaiting the official release of the

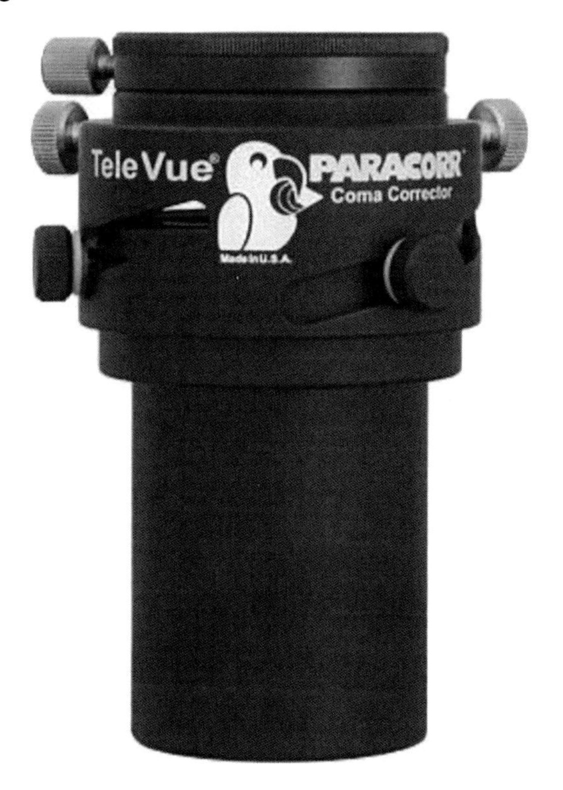

Fig. 9.1. The TeleVue Paracorr (Image credit: TeleVue Optics).

TeleVue Type II Paracorr. The original Paracorr has now been discontinued, though many are available to purchase either new or second hand.

So what are the disadvantages? Well, as already mentioned, you get an additional 15 percent magnification, which results in slightly dimmer views. It also requires about half an inch of in-focus travel and so may not come to focus with some eyepieces. Some Dob owners have had to acquire a shorter profile focuser as a result. At one pound, it's pretty heavy, which may cause some balance problems on lighter Dobs. It's also rather expensive, costing almost as much as a large Nagler eyepiece.

Several other manufacturers have marketed their own versions of the coma corrector, including Lumicon, Baader Planetarium, AstroTech, and SkyWatcher. One of the most popular is the Baader Multipurpose Coma Corrector (MPCC). Like the TeleVue Paracorr, the MPCC is optimized to correct coma in f/4.5 'scopes but does a good job in all 'scopes used visually between f/3.7 and f/5 (and photographically in the range f/4 to f/6).

Fig. 9.2. The Baader Planetarium MPCC coma corrector (Image credit: First Light Optics).

One additional advantage of the Baader unit is that it does not increase the magnification of the system (Fig. 9.2).

For best on-axis performance, all coma correctors need to be located at the correct distance from the focal plane of the eyepiece. The threaded spacer rings on the MPCC makes accurate location simple, with perfect results every time. To attach and properly locate the MPCC, only two or three adapters are probably all you'll need. For instance, you want to use the MPCC with 1.25 inch eyepieces; the only additional components required are the T2-15 reducer/eyepiece holder plus a couple of spacer rings. What you get is a compact and lightweight (considerably less than the Paracorr) coma-correcting assembly that fits snugly inside a 2-inch focuser.

This setup all but guarantees focus in almost any 'scope, including some models with very limited back-focus. The T2-15 eyepiece holder has a captive bronze compression ring with two large knurled screws that firmly clamp down even the largest eyepieces. For short focal length eyepieces, where coma correction is not needed, you can simply remove the MPCC, and the T2-15 goes back to being a high-quality 2-inch/1¼-inch reducer. Used like this, the T2-15 provides 37 mm of spacing. The MPCC screws directly into large, 2-inch eyepieces via its 48-mm filter threads and allows the MPCC to be accurately positioned, providing the optimum spacing for high-quality views. This configuration also enables the MPCC to work without needing any additional back-focus.

In this configuration, the eyepiece will actually come to focus about 10 mm further out than it does without the MPCC. This is a great feature of the Baader coma corrector. More eyepieces come to focus with a MPCC than with a Paracorr. It's also significantly lighter than the TeleVue unit. One caveat that has been reported about the Baader MPCC is that it can introduce more spherical aberration into the optical train than the Paracorr does, which can possibly compromise high magnification views. All said, if you have a fast Dob, a coma corrector is sure to be on the top of your accessory wish list.

Eyepiece Considerations

Eyepieces are the smallest accessories that come with your new telescope, and so you'd be inclined to think that they don't matter much. In short, nothing could be further from the truth. Anyone worth his or her salt in this hobby will tell you that eyepieces play an important role in determining instrument performance.

Eyepieces are big business in the astronomy community. Each astronomer needs at least three of them, and it is common to have many more. Multiply that by the number of amateur astronomers out there, and you see what we mean. Have you any idea how long some online threads run discussing the minutiae of a topical eyepiece? It beggars belief! That said, it is understandable how some folks go to the ends of Earth and back to get their dream eyepieces. These "peep holes" on the sky are sought after and talked about the world over.

And it's a mixture of the old and the new. Some of the simpler designs, employing just a few glass elements, would look right at home in a Victorian shop front. Other designs, enjoying ten or more elements, are so far removed from that era that they look more like grenades than eyepieces – and some cost considerably more than the 'scopes they are used in!

That said, it's not at all necessary to have a dozen eyepieces in order to squeeze the very best performance out of your Dobsonian telescope. Indeed, three carefully selected oculars and a good Barlow lens – which amplifies the magnification by a factor of typically two to up to five times – is often all that's required. But which eyepieces are right for you?

Eyepiece Basics

The magnification of any eyepiece can be found by dividing the telescope's focal length by the focal length of the eyepiece. In general, an eyepiece delivering a "low" power of about 2× per centimeter of aperture is usually best for making broad sweeps over wide fields of view. For zooming in on objects such as galaxies or nebulae with low surface brightness, an eyepiece delivering a magnification of between 5× and 8× per centimeter of aperture yields comfortable, "medium" power views.

For the Moon, bright planets, close double stars, and small planetary nebulae, high magnifications are often necessary. For this purpose, an eyepiece giving 15× to 20× per centimeter of aperture can be pressed into service. Higher magnifications can sometimes be used when the seeing is exceptionally good, but nights like these are very much the exception rather than the rule, especially in the British Isles. In general, seasoned observers use the lowest magnification that enables them to see all of the details an image can yield. Amplifying the image beyond this point will only result in a larger image scale with little or no improvement in detail.

There is also a limit to how low a power you can use with your Dob. This is because an eyepiece's field of view is limited by the diameter of the light shaft that exits an eyepiece – the so-called exit pupil. The diameter of the exit pupil can be calculated by dividing the focal length of the eyepiece by the focal ratio (f-number) of the telescope. Alternatively, it is the diameter of the primary mirror divided by the magnification of the eyepiece. If this light shaft exceeds about 7 mm – the size of a fully dilated pupil in younger people – then your eye will simply not be able to make use of all the light collected by the telescope. Worse still, the shadow of the secondary mirror will rear its ugly head, reducing image contrast. As a general rule of thumb, the longest focal length eyepieces your obstructed 'scope can usefully employ is found by multiplying the focal ratio of your 'scope (see the glossary in this book) by 7. Thus, in a f/5 Dob, never use an eyepiece with a focal length longer than about 35 mm (5×7). To do so, would give a lower image scale, but at the expense of losing light. As you get older, the muscles controlling pupil dilation and contraction get less agile, with the result that even smaller exit pupils are the norm, requiring still shorter focal length eyepieces.

Going Wide

Low magnifications always deliver wider fields of view, but just how much sky your eyepiece can take in also depends on its apparent field of view. A simple, albeit empirical, formula to use when calculating your actual field of view in angular degrees is to divide the eyepiece's apparent field by its magnification. Thus, a 30-mm eyepiece with an apparent field of view of 50 degrees yields a magnification of 50× with a 30-cm/f5 Dob and so provides an actual field of view in the telescope of 50/50 = 1 degree.

Fig. 9.3. Three carefully selected eyepieces and a Barlow can meet nearly every requirement the telescope user has (Image by the author).

In contrast, a 30-mm ultra-wide-angle eyepiece, with a 100 degree field, gives a whopping 2.0 degree field at the same magnification. It's not easy to see what the difference between these two eyepieces are on paper, but one look through both will convince you that the latter view – a truly "spacewalk experience" – takes in an area of sky four times bigger than the former! Wide-angle eyepieces are popular in long and medium focal lengths. The former are used to obtain the widest true field the telescope can offer for low magnification sweeps, while the latter have proven very effective when observing fairly large-sized deep sky objects at close range (Fig. 9.3).

Today's eyepieces vary enormously in the apparent fields of view they offer. Anything from 25 to 100 degrees can be encountered. There was once a time when only a few manufacturers were selling expensive, wide-angle eyepieces, but in recent years the market has been flooded with budget-priced, mostly 2-inch barrel diameter models that boast enormous fields of view. But how good are they?

In some ways, you get what you pay for. The bottom line is that while there are now many wide-angle eyepieces on the market, they differ markedly from each other in terms of the extent to which they correct the field of view. For one thing, chunky wide-angle eyepieces possessing multiple glass elements must be fully multicoated in order to minimize any internal reflections that would otherwise plague an image. Thankfully, this has now become an industry standard.

By far the most annoying aspect of less expensive, wide-angle eyepiece views is the loss of definition and image distortion as the object is placed, off-axis, away from the center of the field, especially when using short focal ratio (from f/6 to f/4) instruments. These eyepieces, while presenting nice, pinpoint star images in the inner 50 percent of the field, say, generally display increasingly inferior images – most often astigmatism and lateral color – as one moves the stars to the edge of the field. Yet, when you use the same eyepieces in an f/8 'scope, even inexpensive models can perform admirably. That said, the finest (and most expensive) wide-angle eyepieces allow you to have your cake and eat it, so to speak, with expansive fields of view and tack-sharp star images all the way to the edge! There are now a variety of wide-angle eyepieces that give well corrected low power views with fast Dobs on the market. Check out the ranges offered by TeleVue, Meade, Baader, Orion (USA), Vixen, Sky Watcher, Explore Scientific, and Pentax to see what we mean.

What's Your Comfort Zone?

Eyepiece enthusiasts have never had it so good . There was once a time when there was little choice available to the dedicated observer in regard to the types of eyepieces he or she could employ. Those eyepieces – Plossls and orthoscopics, for example – although comfortable to use in long focal lengths – become a nuisance to use at short focal lengths because one has to place one's eye right up to the glass to access the entire field of view. In the worst cases, the eye has to be placed just a millimeter or two above the eye lens of the ocular to see the full field. In technical jargon, we say that these eyepieces offer very little "eye relief."

Specifically, eye relief measures how close your eye has to be to the top of the eyepiece to take in the entire telescopic field. In the last few years there has been a concerted effort made by telescope manufacturers to offer so-called long eye relief eyepieces to make viewing as comfortable as possible. As a general rule, shorter focal length eyepieces have less eye relief than their longer focal length counterparts, meaning that you've got to place your eyeball closer to the lens. Most observers agree that between 15- and 20-mm eye relief give the most comfortable views, and observers who wear glasses should elect to use eyepieces with at least 15 mm (and preferably 20 mm) of eye relief.

It's important to remember that there is much debate on the amateur forums about which eyepieces are best suited to viewing particular celestial objects. Some enthusiasts still turn to minimalist designs such as orthoscopics and Plossls to drink up the brightest views of faint deep sky objects. The reduction in the number of air-to-glass surfaces in these eyepieces helps to ensure the greatest possible transmission of light. Other Dob owners use the restricted fields of view of these "old school" eyepieces to keep objects centered inside the diffraction-limited fields of view of their 'scope. Recall that the faster the f-ratio of the Dob, the smaller its diffraction-limited field becomes. Thus, in this capacity, simple eyepieces such as monocentrics, Plossls, and orthoscopics are less "old school" and more contemporary than we have given them credit for.

Zoom Away

Why buy three or four eyepieces when you can purchase a zoom model that can vary its focal length from 8 to 24 mm, or 6.5 to 19 mm, for example? On paper, there's no denying that the idea is very attractive, For one thing, you obviously remove the need to keep interchanging eyepieces. This is particularly handy if you need to keep an object firmly within the field of view or when demonstrating an object in public outreach events. The execution of a successful zoom eyepiece design is another matter, however.

All too often, its convenience is offset by its mediocre performance. Many of the standard models (such as the Meade/Vixen/TeleVue 8–24-mm zoom) tend to have very restricted fields of view, often as low as 35 or 40 degrees at the lowest power, as well as requiring the need to re-focus at each magnification setting. What's more, the extra glass elements employed in the "early" zooms tended to serve up images that were noticeably softer than dedicated, fixed focal length eyepieces. That said, in recent years, it looks like some manufacturers have raised their game. Now zoom eyepieces manufactured by Baader, Leica, TeleVue, and Pentax show real signs of competing with fixed focal length eyepieces in terms of the image quality they deliver. For lunar and planetary enthusiasts, TeleVue Nagler zooms (particularly the 3–6 mm) come highly recommended. The latter's five element design has proven to be every bit as good as the majority of fixed focal length counterparts. As ever, expect to pay top dollar prices for the best models.

Observing the Moon, the Planets, and Double Stars

While observers seem to have reached a consensus on the best eyepieces to use in low and medium power applications, lively debates are still being held in many on-line chat rooms about what features a good planetary eyepiece should possess, apart from perhaps having excellent on-axis sharpness with little chromatic aberration. And it's easy to see why. Do you embrace simple designs with little eye relief and restricted fields of view or do you put your trust in the promises of the modern age, with its rich offerings of wide-angle eyepieces with generous eye relief to view the Moon and brighter planets at high magnification? All other things being equal, a wide field of view is an obvious bonus, but many seasoned observers believe that there's a definite trade-off in performance when using wide-angle eyepieces for high magnification lunar and planetary work.

For one thing, wide-angle eyepieces tend to be heavier and have more glass components than traditional planetary eyepieces from the monocentric, Plossl, and orthoscopic genre. The more air-to-glass surfaces an eyepiece contains the more likely it will suffer from ghosting (internal reflections). This, together with the fact that a little light is lost when traversing each lens, it is easy to see how an eight-element wide-angle eyepiece might give quite a different image to a simple orthoscopic, say, with its four glass elements. In particular, many experienced observers note that the image of planets in complex, wide-angle models tends to be slightly yellower or "warmer" than the "cooler," more spectrally neutral images delivered by simpler designs.

As you may have gathered, eyepieces are, and will continue to be, hot topics with amateur astronomers, but it's up to you to decide which models are best for you. Because the market has so many competitors, there's bound to be a level of subjectivity when it comes to finally deciding the eyepiece set you finally settle on. Even expensive eyepieces can be had for modest sums if purchased on the used market. If at all possible, try before you buy, but if not, good quality oculars can be re-sold for reasonable prices so you can always get some money back from your investment. Good luck with your eyepiece shopping adventures!

Barlows and Image Amplifiers

There is yet another tried and trusted way of achieving high powers and even improving the performance of your favorite eyepieces – use a Barlow lens! This simple device has been around for the best part of two centuries. It was the English mathematician and engineer Peter Barlow (1776–1862) who first hit on the idea of introducing a negative (concave) achromatic doublet ahead of the eyepiece to artificially increase the effective focal length of any telescope by a factor of two or three times. Working with George Dollond, the first true Barlow lens was fashioned in 1833. Yet, until recent times, Barlows have endured a love-hate relationship with amateur astronomers. The units that used to accompany cheap department store 'scopes did much to dispel their real potential as power boosters for many decades. That was no doubt because of their poor optical and mechanical quality. However, recent advances in manufacturing technology has now made owning a quality Barlow lens a truly worthwhile investment, especially considering their great versatility and modest price tags.

Not only will a good Barlow lens double or triple the power of your eyepiece, it will reduce any aberrations inherent to the eyepiece by creating a gentler sloping light cone that is more faithfully reconstructed by the eyepiece-eye combination. It will also flatten the field, helping to reduce edge of field distortions in short focal length refractors. So-so quality eyepieces will perform noticeably better when used with a high-quality Barlow lens, especially when the object is placed at the edge of the field. Barlows will *not* increase the telescope's depth of focus, however, as is commonly believed. That's because the Barlow lens, when used in the usual way, is placed in the focuser ahead of the eyepiece and moves with respect to the primary and secondary mirrors. If the Barlow were placed *ahead of the focuser* and *fixed* in position relative to the telescope's primary mirror, then it would increase depth of focus. That would certainly be a neat project for an amateur telescope maker!

Barlows come in two varieties: achromatic doublets and so-called apochromatic, or ED, models employing three or more lenses. You might think that having the name ED or apo on the Barlow necessarily implies better performance. However, that's just marketing hype. Extensive tests on a number of Barlows can be summarized as follows: well-made two-element achromatic Barlows (2× and 3×) consistently outperform the shorter, "cuter" apo Barlows when it comes to overall image quality and light throughput.

That's not to say that your cherished Celstron Ultima or Klee (two highly respected "shorty" Barlows) is not a very good performer. Far from it! It's just that a three-element ED Barlow is overkill from an optical standpoint. The best Barlow lenses are fully multicoated achromatic doublets. In addition, they are internally baffled to minimize stray light. All Barlow lenses increase (to a greater or lesser degree) the eye relief of eyepieces used with it. Viewing through a 6-mm Plossl, for example, without a Barlow is tricky but becomes fairly easy when used in conjunction with one. A shorter, triplet Barlow will look less conspicuous on your Dob, but it will almost certainly increase the eye relief more than a traditional doublet Barlow. That isn't a problem for short focal length eyepieces, which have fairly modest eye relief to start with, but it may generate uncomfortably long eye relief if you intend using it with low-power, longer focal length eyepieces. Both TeleVue and Astro Physics make some of the best. The Tal 2× and 3× models are excellent performers for their modest prices. The Zeiss 2× Barlow is even better, but it'll cost you more ($495).

Although good Barlows work really well, they have one potentially annoying drawback, especially when used with an eyepiece that already has the optimum amount of eye relief. Using a Barlow forces the observer to place his or her eye into a new position, which may or may not help. In particular, Barlows work fantastically with traditional eyepieces such as orthoscopics and Plossls, which have little eye relief to start with, but when coupled with eyepieces that already have generous (> 15-mm) eye relief they can extend it too much. What is desirable in this situation is an amplifier that preserves the native eye relief of your eyepiece while still giving you the magnification increase you desire.

Enter the Power Mate, a cleverly designed device introduced by TeleVue over a decade ago. Like an ordinary Barlow, it consists of a doublet tele-negative element to increase the effective focal length. Al Nagler then had the presence of mind to insert an additional tele-positive element (also a doublet) that essentially reconfigures the light path so that it matches the eye relief of the original eyepiece. These gadgets really do add nothing but raw magnification to your optical system and are an excellent way to achieve high power in fast reflectors (short focal ratio). Meade also produces a very similar line of image amplifiers to the Power Mates. Called Telextenders, they come in 2×, 3×, and 5× (all in 1.25-inch format) as well as 2-inch (2×) version.

The main astronomy forums are constantly abuzz with arguments about the pros and cons of Barlow lenses, with some innocently posed questions quickly escalating into highly technical and heated debates. There are a few reasons, partly rational, partly aesthetic, why lots of amateurs shy away from Barlows. They stick out a mile from the focuser, creating, at least in the minds of some folks, an aesthetically displeasing disposition. Others maintain that adding more glass into the optical train reduces the "punch" of the image. As has been said already, this is based mostly on incorrect assumptions. Diehard observers who doggedly refuse to use a Barlow would do well to remember that most high-power eyepieces with short focal lengths already have a Barlow built in! How else do you think modern eyepieces can easily achieve ridiculously short focal lengths (2 mm to 4 mm) while still maintaining comfortable eye relief?

Filters for Deep Sky Work

The Dobsonian revolution has, above all things, allowed ordinary people to enjoy the advantages of very large aperture to probe the depths of the deep sky. Big Dobs are light buckets, capable of capturing scintillating views of star clusters, galaxies, and emission nebulae of all kinds. So why

go to all the trouble of acquiring a large aperture Dob only to stick a filter in the eyepiece to cut off some of that light? That's nuts, right?

One would have thought that any kind of filter – with its inherent light loss – would be counterproductive for probing dim nebulae of the deep sky. The truth is this is one province of observational astronomy where filters have unequivocally proven their worth. These handy devices come in three varieties: light pollution reduction (LPR), ultrahigh contrast (UHC), and line filters (OIII, hydrogen beta). LPR filters block quite a bit of unwanted artificial light (such as the ubiquitous sodium and mercury). Still, while they work very well in photography, they fail to excite many deep sky observers because they just don't seem to be aggressive enough at "pulling" faint nebulae out of the background sky. Nonetheless, LPRs can work quite well at low and medium power with smaller Dobs up to 8 inches in aperture from moderately dark skies.

Much better again are the UHC filters, which pass a narrow bandwidth of light – typically 25 nm, centered on the most prominent visible radiations issuing from the myriad emission nebulae scattered across the sky. These include hydrogen alpha (Hα) and beta (Hβ) radiation – both useful for diffuse nebulae, as well as the light emanating from doubly ionized oxygen (OIII), a wavelength at which planetary nebulae shine brightly. While these filters typically dim stars by about one magnitude, they dramatically darken the sky. Yet despite the light loss, you *can* see faint emission and planetary nebulae better. UHC filters are especially useful for smaller telescopes. One of the most memorable and enduring views this author enjoyed of the Cygnus Veil was using a UHC filter on 6-inch f/5 at 30×.

Arguably, the most effective deep sky filters has got to be the venerable OIII – so-called because it only passes a thin waveband (10 nm) of light centered on a pair of lines emitted by doubly ionized oxygen (an oxygen atom that has lost two electrons). Since this radiation is especially enriched in planetary nebulae, OIIIs are the single best filter to use when studying these small, ghostly glows. Field stars are dimmed even more with OIII filters than with the UHCs, and for this reason they are best used with 'scopes larger than about 6 inches (15 cm). There's quite a bit of variation in the quality of these filters which come in a 1.25- or 2-inch fit version. For example, how well polished are their surfaces? If the interference film has a roughness that amounts to more than ¼ wave, then it could noticeably affect high-power views. On the contrary, if these filters are polished to an accuracy of less than ¼ wave, they are said to be diffraction limited and thus will not degrade the image. Both UHC and OIII filters work best at low and medium power applications with relatively short focal ratio (f/7 or lower), so don't be tempted to crank up the power too much while using them.

The hydrogen beta (Hβ) filter has an even narrower bandwidth (8 nm), centered on the Hβ line at 486 nm. This filter is far less versatile than either the OIII or the UHC and can only be used to good effect on a very limited range of objects such as the Horsehead Nebula in Orion or the California Nebula in Perseus. Unfortunately, Hβ filters often gather more dust in an eyepiece box than starlight.

Battling Against Dew

The vast majority of amateur astronomers are bound to encounter dew problems at some stage in their career. In case you didn't know, when the mirror in your 'scope dews up, you rapidly lose contrast on deep sky objects, and lunar and planetary images noticeably deteriorate. Nor do you need to see dew on your mirrors before it has an insidious effect on the telescopic image. The late Walter Scott Houston commented that microscopic amounts of dew on his 'scope's objective – below that which could be discerned by the human eye – lost a whole magnitude of light grasp. If you're a frequent observer, battling with dew is a necessary part of life, so it pays to know how it forms in order for you to prevent it from ruining your observing programs.

Dew forms when the outer surface of the mirror cools down below a temperature known as the dew point. The mirror warms the air immediately above by a mixture of convection and radiation. At first, convection dominates, but as the mirror approaches the surrounding air temperature, radiative cooling begins to dominate. When the mirror temperature falls below that of the ambient air, the air starts to convect heat – and any water vapor contained therein – back onto the mirror, and you get dew build up.

It follows that since smaller mirrors cool off faster than larger ones, the former are more prone to dewing than the latter. Some amateurs use a dew shield – the longer the better – to slow radiative cooling. But if you like observing for prolonged periods of time, this approach is still no guarantee against dew forming on your optics. The only real solution is the addition of a small amount of heat; enter the dew heater, which we'll discuss shortly. You may have noticed that on windy nights, dew hardly ever forms. That's because moving air counteracts the effects of radiation-induced cooling, so it keeps the mirror above the dew point. Ironically, although it helps cool down your primary mirror, a fan is also an effective dew prevention device. In damp climates the secondary mirror is far more prone to dewing up in the field than the better insulated primary sitting at the bottom of the tube.

First off, under no circumstances should you attempt to wipe off the dew from the mirror surface. That kind of behavior will, sooner than later, damage its delicate coatings. As intimated above, a dew shield can only be said to delay the onset of dew rather than as a preventative measure. A dew heater consists of a strip of electrical resistors shrouded in cloth that wraps around the outside of the mirror and heats the corrector plate to just above ambient temperature, preventing dew from forming. The heaters for secondary mirrors are placed on the back side of the secondary and can be held in place by using the polyester batting found inside the secondary mirror holder. There's lots of information available online about how to make your own dew heater. Failing that, companies such as Kendrick Astro Instruments have a very extensive range of dew prevention systems that can be purchased at modest cost (Figs. 9.4 and 9.5).

Fig. 9.4. A dew heater for a secondary mirror (Image credit: Kendrick Astro Instruments).

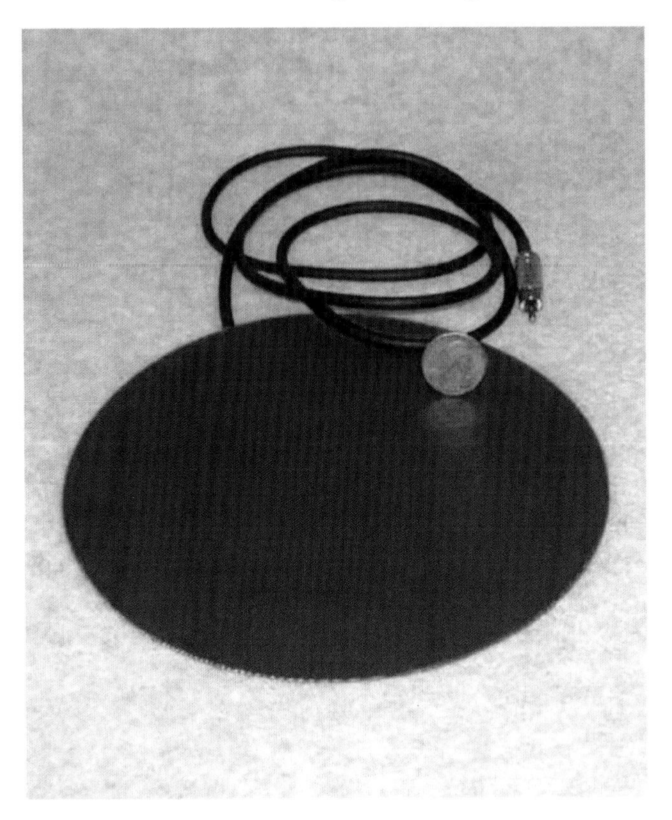

Fig. 9.5. A heating pad for a primary mirror (Image credit: Kendrick Astro Instruments).

Getting Your Dob to Track

As we have said more than a few times in this book, a well-made Dobsonian telescope is an excellent instrument for exploring the night sky. When properly designed, the mount upon which the 'scope glides can be very stable, with smooth motions. What's more, it can be very compact and eminently transportable. However, to some observers at least, the Dob has one major drawback – it has no motor drive. You have to push it around the sky. And for some, that can get to be a hassle. But what if you were able to perch your Dob on an equatorial platform and experience instant motorized tracking! Wherever you point, you're tracking, smoothly and precisely.

Equatorial platforms for Dobs have been around for quite some time now. The principle upon which they are based is fairly easy to grasp.

Typically the platform consists of two parts: a wooden ground board and a pivoting table, to which the board is attached. The pivot axis is aligned at an angle equal to your latitude (or the altitude of the Pole Star above the horizon). Typically, commercial equatorial platforms can accommodate latitudes from 0 to 55 degrees north or south.

A drawback of most of these platforms is that they are only able to continuously track the sky from anywhere between 30 min and 90 min before having to reset the base all over again. When this happens, you've got to turn off the stepper motor and pull up on the reset handle. This will tip the platform to the east, making it ready for another hour of tracking. Although this is a very simple and fast operation, you will lose whatever happened to be in the eyepiece at that time.

Several companies market these tracking mounts. Equatorial Platforms, a company founded by Tom Osypowski, has two highly regarded models. The more economical, apple-plywood constructed platform ($975 plus shipping and crating) can hold Dobs up to 25 inches in diameter and track continuously for 75 min. When equipped with dual-axis motor drives, the platform is also more than capable in engaging in advanced astrophotography. Needless to say, reports on this model are invariably positive. The whole unit runs smoothly and quietly on a single 9 v battery (Fig. 9.6).

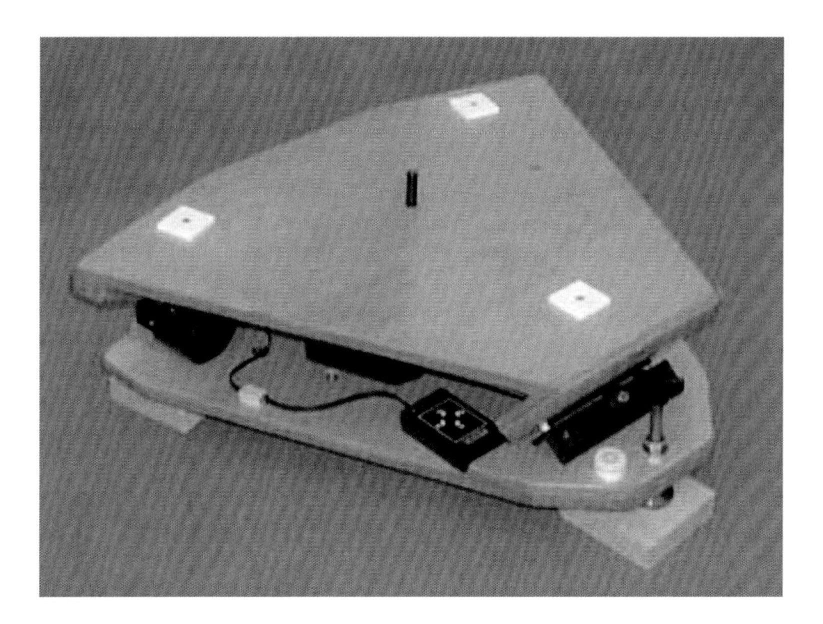

Fig. 9.6. A compact equatorial platform (Image credit: Tom Osypowski).

Fig. 9.7. A 12.5-inch truss tube Dob astride an equatorial platform (Image credit: Ron Klein).

Equatorial Platforms has also launched a more robust, aluminum-based tracking platform that can accommodate 'scopes of between 16 and 32 inches in aperture. They work like a dream but are more costly than the wooden models described above ($2,000 and upwards) (Fig. 9.7).

An exciting new product developed by the Arizona-based company StellarCAT has made a lot of Dob owners happy. We're talking, of course, about the innovative ServoCAT GoTo system for Dobs. Easy to install on just about any Dob, ServoCAT offers powerful GoTo capability. Selecting from a number of menus, including Sky Commander (9,000 objects) and Argo Navis (30,000 objects), the ServoCAT slews at up to seven (five for the biggest Dobs) degrees per second (one of its eight slew speeds) to any object above the horizon that is contained within its database. Like the more conventional GoTo systems, such as Meade's Autostar system, it can do local "syncs" to refine pointing accuracy and also offers the user a "spiral search" option whereby you can ask the ServoCAT to search the surrounding hinterland. StellarCAT offers its GoTo systems in "junior" form to cater to 'scopes with apertures of between 8 and 15 inches (ServoCAT Jnr. $1,329) or the more established ServoCAT ($1,479 to $1,659, depending on the size of your Dob) system designed for the biggest commercial Dobs.

So what's the advantages of using a ServoCAT in comparison to a tracking platform? For one thing, you won't have to carry it around with you, so it cuts down on gear to transport and set up. Equatorial tracking platforms also raise the 'scope further off the ground, but not so with ServoCAT. That may not be a big issue for smaller Dob users, but when you get to the 20-inch monsters having a way to set up something on the ground can mean the difference between having to use a step ladder and not.

The global economic recession of 2010 took its toll of victims in this market, too. Several well-established manufacturers of high-quality Dobsonian platforms, such as Johnsonian Engineering and Round Table Platforms, have gone out of business. This already niche market has also suffered at the hands of the new high-tech Dobs – big guns with motorized and even GoTo technology built in.

Global recession aside, in 2010 Sky Watcher pushed ahead with their ambitions to launch their innovative new product, the Flextube Auto Dobs. Building on the success of the innovative flextube truss design described earlier, the company quickly launched the same product with built-in, heavy-duty, dual-axis servo-tracking motors (Fig. 9.8).

The hand controller, which attaches neatly to the side of the 'scope, is used to slew the instrument to almost any position in the sky. The motors are powered off a 12 v DC source. With the tracking mode set to the "off" position, the telescope can be slewed at a choice of slow (32× sidereal), medium (64×), or fast (800×) speeds. When tracking mode is switched on, the user can make adjustments at 1×, 4×, and 8× sidereal rates, which is especially useful when trying to center a target or for imaging applications.

What is particularly remarkable is just how quiet the whole 'scope moves even when slewing at the highest speeds. The automatic mode has a unique and patented override, which enables the user to manually move the telescope to a position and retain the tracking when you get there. This is very useful, as it can often be quicker to move the 'scope manually

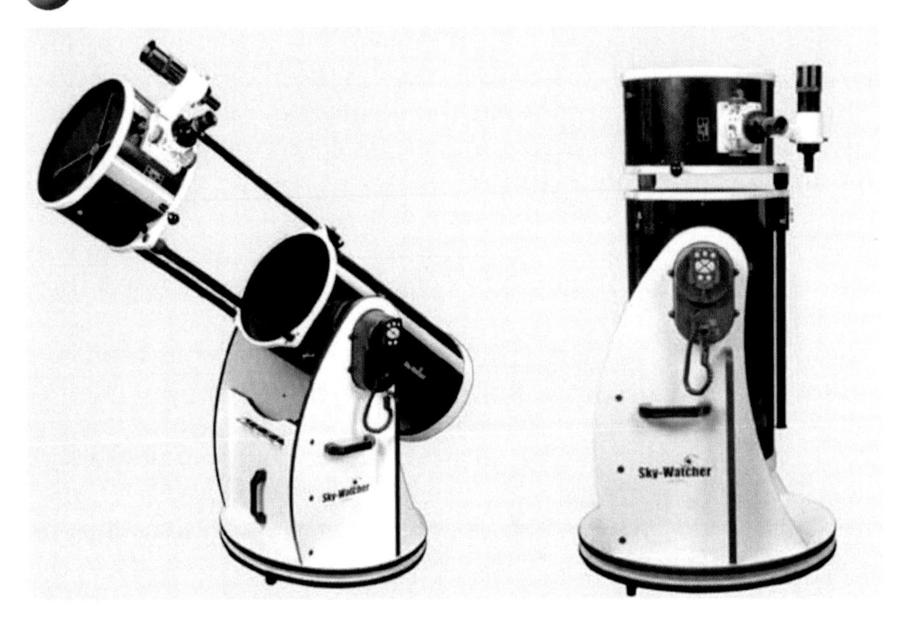

Fig. 9.8. The innovative Sky Watcher Flextube Auto Dobs (Image credit: Optics Star).

into position than to slew it across the sky. The exciting thing about these auto tracking Dobs is that they open up large aperture instruments to astro imagers – a subject we shall be exploring in a later chapter.

Once built-in automatic tracking became a reality for big Dobs, it was only a matter of time before the same technology could be upgraded to full GoTo capability. Enter the Orion SkyQuest XTg series (Fig. 9.9).

Introduced in the summer of 2010, the XTg series (available for its XT8, XT 10, and XT 12 Dobs) offers all of the features seen on the Sky Watcher auto series but with the added ability to automatically go to any one of 42,000 objects stored in its database. You have a choice of aligning the 'scope via "level-North" or using a "two stars" method. Both work well., though they are not always dead center on the object. Not surprisingly, the GoTo capability of the XTg 'scope is considerably more accurate for objects located near the two alignment stars. Tracking is very good. Indeed, it keeps objects in the field of view with a medium power eyepiece for the best part of an hour.

That said, there are a few issues with the 'scope. It feels as if it is not meant to be used manually. There's quite a bit of stiffness, especially when moving the azimuth bearing. Indeed, trying to move the 'scope in

Fig. 9.9. The Orion (USA) SkyQuest XT8g, a GoTo Dob (Image credit: Orion Telescopes)

azimuth while it's pointed high overhead can lead to tipping the base. There is also the "creaking altitude bearing" issue that many have reported while using the 'scope in the field.

That brings us to the end of this chapter on accessories for your Dob. Never has there been a better time to buy into the Dobsonian revolution and all the wonderful accoutrements that have been developed to complement their use in the field. In the next chapter, we'll be taking a look at various ways of adjusting and testing the performance of your Dob under the night sky.

CHAPTER TEN

Adjusting, Maintaining, and Testing Your Dob

So you've got a great big Dobsonian. How do you get the most out of it from an optical standpoint? Are the mirrors aligned properly with your inserted eyepiece? Do you want to evaluate the optical quality of your 'scope? This chapter addresses all these issues and helps you get the maximum performance for your hard-earned investment. Let's begin with the age-old Newtonian chestnut – collimation – and how best to execute a near perfect alignment of your optics.

Collimation simply means that all the optical components making up your telescope are perfectly aligned. So the primary mirror must be in line with the secondary flat mirror, which in turn must be centered in the eyepiece drawtube of your 'scope. Fairly accurate collimation lets your 'scope perform well on low and medium power targets, but for the best high power views, a little extra care in collimation can make the difference between a good and a great image. Accurate collimation becomes extremely important as the F ratio of your primary mirror decreases. For example, a f/8 mirror is far more forgiving to slight mis-collimation compared to its f/4 counterpart.

Many Dobs come with their primary mirrors center-marked with a small paper disk to help the collimating procedure. Some also come equipped with a so-called "collimating eyepiece." Failing that, you can

N. English, *Choosing and Using a Dobsonian Telescope*, Patrick Moore's Practical Astronomy Series 1, DOI 10.1007/978-1-4419-8786-0_10,

make one yourself out of an old 35-mm canister. First cut out the bottom of the canister and then drill a hole about 2–5 mm in diameter in the exact center of the cap. The canister is just the correct diameter and will fit nicely into the focus tube. The hole enables you to look directly down the center of the focus tube, where you should see the following: the focus tube, the secondary mirror along with its holder and clip, the reflection of the main mirror, the spider vanes, and finally, the reflection of the secondary mirror.

To begin with, what is required is that the secondary mirror be perfectly centered underneath the focus tube. If the secondary mirror is offset to the left or right of center, you will need to adjust the spider vane screws on the outside of the telescope tube to bring the mirror assembly back to center. Do this systematically by making sure to unscrew one vane screw while screwing in the other. If the secondary mirror is too far in or out of the tube, you'll need to loosen the three set screws on the secondary mirror assembly and then either screw in or out the center screw that holds the secondary mirror in place. This process should bring the secondary mirror directly underneath the focus tube. You should probably rack the focuser in to the point where the secondary mirror just covers the focus tube opening. In this way, you will be able to make fine adjustments until everything looks symmetric.

The next step is to center the main mirror's reflection on the secondary mirror. To do that, you'll need to identify the three set screws on the center of the spider assembly directly behind the secondary mirror. While looking down the collimating eyepiece, loosen these three screws and tilt the secondary mirror until you are able to see the entire main mirror on the secondary mirror. Lightly tighten the set screws and proceed to make fine adjustments to them until you judge the main mirror's reflection to be perfectly centered on the secondary mirror. After this step you might need to go back and ensure that the secondary mirror is still centered underneath the focus tube. If not, adjust the spider vane screws again and check that the main mirror is still centered on the secondary mirror. Repeat these two steps until everything looks perfect.

The next step is to adjust the main mirror. Here is where a spot on the center of the primary mirror comes in very handy. Two people make this job much easier to accomplish, especially if you have a long tube and can't reach the adjustment screws on the primary while comfortably looking through the focuser.

The main mirror assembly normally has three so-called "set" screws and three "lock" screws. Loosen the lock screws, and then screw in or out one set screw at a time until you get an idea of what each screw does with respect to the center spot. The goal is to adjust the set screws until the center spot is perfectly centered within the collimation eyepiece, which in this case acts rather like a Cheshire eyepiece. At this point your telescope should be fairly well collimated. You can test that out for yourself by aiming the 'scope at some terrestrial subject to see how well it images. Fine tuning is usually done by pointing the telescope at a distant star using an eyepiece that delivers a magnification of between 30× and 40× per inch of aperture.

Polaris is a great target (if you live in the Northern Hemisphere, that is) for a large Dob, as it is so close to the celestial pole that it hardly moves in the field of view of the 'scope the entire time the collimation is being tweaked. Be sure that the star is perfectly centered in the eyepiece and then rack the focuser in and out of focus. When the star is out of focus, you should be able to see the shadow of the secondary assembly and the spider vanes. The shadow of the secondary assembly should be centered within the light circle produced by the star. If it's not, you will need to adjust the main mirror set screws in the direction of the asymmetry. The adjustments required at this point will be very minute, so be careful not to overdo them. Laser collimators are also widely available to assist in the collimation process. If you have a few dollars to spare, buy one. They are less hassle than using a regular Cheshire eyepiece, especially when collimating larger telescopes because you won't need to move back and forth between the eyepiece and the adjustment screws at the back of the 'scope every time you make an adjustment. Even the simpler models have variable brightness settings, so you can tweak the laser for better visibility in daytime or twilight. Many laser collimators include a removable adapter for use in both 1.25- and 2-inch focusers.

Cleaning Mirrors

A mirror with a substantial layer of dust deposited on it will still perform so well that the average person wouldn't be able to tell the difference between the images it serves up and that of a pristine, dust-free mirror. So before embarking on a mirror-cleaning project, make sure it is worth the trouble. First rinse the mirror under the hose to wash away sand and grit. Note that your mirror will be much happier laying face up on damp grass than in the kitchen sink. Just watch out for the Sun's reflection!

Next immerse the mirror in diluted dishwashing detergent in warm water in a plastic pan. Wipe lightly with paper towels to remove grit without scratching the soft aluminum. If the coating is new you may want to pre-soak the paper towels. Wash and rinse with warm water. Final drying is most critical. All drops must be absorbed into paper. Blow drying will leave drop marks. Use caution, because stiff paper may scratch the coating and soft paper can leave lint. Blotter paper may be pulled across the mirror for drying. If the coating is new do not press down as you pull the paper. Some observers have removed the water from their mirrors with 70 percent rubbing alcohol and dried it with ordinary toilet tissue. All things considered, if your Dob has a dust cover and dew is kept from the mirror, then it really should not need cleaning for several years.

Star Testing Your Dob

As we saw in Chap. 2, there are five Seidel aberrations that the star test can pick up on. These are astigmatism, coma, spherical aberration, field curvature, and distortion. Star testing is the most inexpensive and most sensitive way to evaluate how good your Dob is from an optical standpoint. As the guru of star testing Harold Suiter points out in his book, *Star Testing Astronomical Telescopes*, a skilled interpreter can diagnose all sorts of problems that might otherwise slip through the net during normal use. In fact these tests are so sensitive that no telescope ever built can pass them with 100 percent success! By implication, it's a reminder that no telescope is absolutely perfect. If you detect tiny imperfections, it will probably deflate your ego more than your optics. More than a few amateurs avoid doing detailed star tests because they don't want to get upset dwelling on aberrational minutiae. It's hard to blame them!

To perform the star test, wait until the seeing is good to excellent and take extra care to ensure that your optics are well collimated and have cooled down long enough for the images to settle in the eyepiece. As a general rule, it is inadvisable to make definite conclusions about the quality of your optics by performing a single star test. You need to do it several times to render an accurate assessment.

Common Defects

Let's now look at each type of defect in turn. In all cases, examine a bright star that is centrally placed in the field of view. Polaris is a good test star, as it is very near the north celestial pole and so moves little during the course of a typical star testing program. Use an eyepiece that delivers a magnification somewhere between 30× and 50× per inch of aperture.

ASTIGMATISM: Point your 'scope at a bright star and focus it as sharply as possible. Now move the star *ever so slightly* outside focus. The Airy disk should remain round. If it appears egg shaped, rack the 'scope *ever so slightly* inside focus. Did the egg flip in shape through 90 degrees? If so, you've got some astigmatism. Most short 'scopes show some, especially when tested at very high powers. Finding a slight amount of this aberration at high powers will not appreciably degrade your images.

COMA: As explained earlier in this book, off-axis coma is ever present in fast Dobs (< f/5). But to see it on-axis is not encouraging. To detect it, focus the bright star as well as possible once again. Is the light emanating from the star symmetric in all directions? Or can you see ripples, like the gills of a fish (or a comet tail, if you prefer), emanating from one side of the star?

SPHERICAL ABERRATION: Examine the intra- and extra-focal image of the bright star again. Are the rings noticeably easier to see on one side of focus compared with the other? In particular, is the outer ring brighter on one side of focus relative to the other? If yes, then you're probably picking up some spherical aberration. Spherical aberration comes in two flavors: under correction or over correction. If the outer ring is brighter inside focus than outside, then you have some under correction. The reverse is true if your 'scope displays over correction. It is telling that most commercial Dobs tested tended to be slightly or moderately under corrected. There's little you can do to ameliorate this defect from your 'scope, apart from getting the mirror refigured.

It is important to remember that a mirror that has not had time to sufficiently cool down will exhibit signs of over correction, so observing this may reflect a thermal issue than any inherent problem with mirror. Conversely, if a mirror is under corrected to start with, it may exhibit superior images if temperatures continue to fall during a typical observing session.

FIELD CURVATURE: Take a low power eyepiece with good edge of field correction and insert into your 'scope diagonal. Center a bright star in the field of view and focus the image as sharply as possible. Now slowly move the star to the edge of the field and examine how the image of the star changes. When the star is at the edge of the field, do you have to refocus it slightly to get the sharpest image? If so, your 'scope is probably showing some field curvature.

DISTORTION: Not an aberration in the same way as the other four. It usually is seen when using wide-angle eyepieces. It comes in two forms: *pincushion* (positive distortion) and *barrel* (negative distortion). These are best seen during daylight hours by pointing your 'scope at a flat roof and looking for bending of the image near the edges of the field. These defects often arise in the eyepiece rather than your telescope's objective lens, and while they can be slightly distracting during critical daylight tests, they can't be seen during observations conducted at night. So if you're an astronomer, distortion matters little.

Other Defects to Look Out for

Sometimes mirrors, despite excellent seeing conditions, produced barely perceptible (sometimes totally invisible) diffraction rings. Instead of seeing nice concentric rings either side of focus, you see a dog biscuit-like pattern. This condition indicates a rough optical surface caused by inadequate polishing. You could have a Dob that gives a lovely 1/10 wave figure, but in the presence of a significant amount of surface roughness, the optical punch would be more equivalent to standard ¼ wave optics. Fortunately, there is far less of this in commercial Dobs of late; even inexpensive models now seem to have well polished optics. Sometimes, the primary mirror is held too rigidly in its cell, and, as a result, pinching can occur. That can be identified by odd triangular diffraction patterns in the in-focus star image.

Mirrors are also prone to so-called local or zonal errors, that is, localized defects that arise during the figuring and polishing of optical mirrors. Luckily these zonal defects are most likely to occur at the center of the mirror and so are of little consequence, since they are in the shadow of the secondary. Zonal defects are much more serious if they occur on the outer edges of the mirror, because a greater proportion of the surface area resides in the outermost part of the mirror. The most common zonal errors are *turned edges,* which occur when the edge does not end abruptly but curls over gradually, starting from about 80 percent of the way out from the center of the mirror. Turned edges have the potential to seriously degrade an image. It's fairly easy to pick up in a star test. Inside focus, the rings will appear soft, as if puffed up, while outside focus they take on a very definite hardening, as if someone had highlighted them. The reader is best referred to Suiter's excellent book on star testing (cited in the reference section) for further details on how to pick this up.

The Ronchi Test

Another good way to evaluate the optics of your 'scope is to perform a Ronchi test. Though not as sensitive as the star test, the Ronchi test involves replacing a diffraction grating with 100 lines per inch in place of the eyepiece. Before carrying out the test, ensure that the telescope is accurately collimated and thermally acclimated, as any misalignment will give spurious results. The tests should be carried out at night. Point your telescope at the Pole Star if you do not have a right ascension (RA) drive fitted. If you do have an RA drive fitted, any bright star will do.

Position the star in the center of your field of view with a low power eyepiece and then replace the eyepiece with your Ronchi band adaptor. Focus as you would normally, and you should see a series of black and white bands beginning to appear. Adjust your focuser until about three or four bands are visible and ensure you are within the focal point and not outside it. With a perfect telescope, a Ronchi test should give you perfectly straight black and white bands arranged parallel to each other with no deformities at all (see Fig. 10.1 below).

This test is actually very sensitive and can detect minute errors that have no appreciable effect on images. However, some errors can appear small on the test but have a marked effect on image quality, in particular so-called turned down edges. Don't get paranoid if small errors are present.

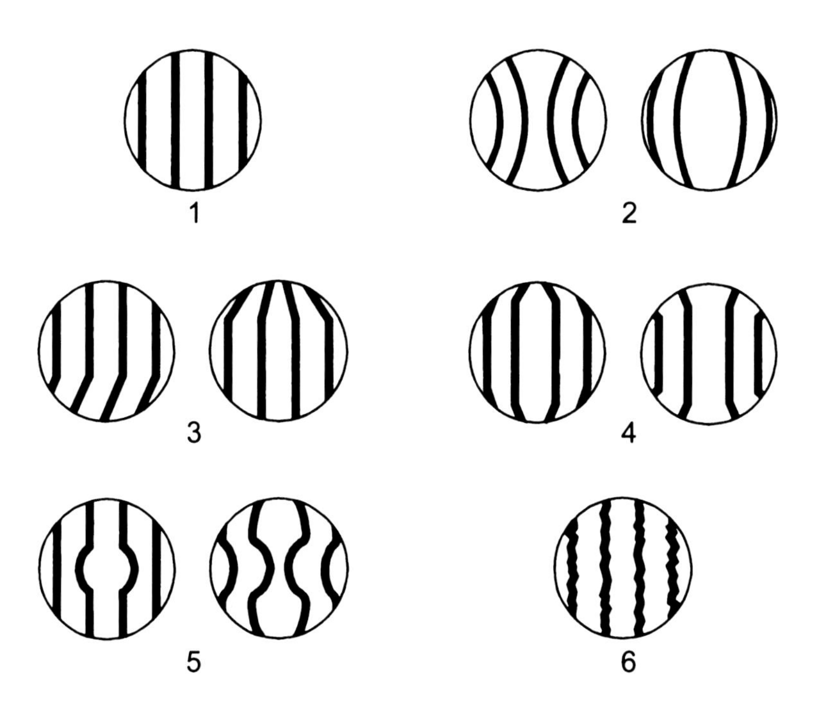

Fig. 10.1. Some Ronchi test results: 1. Perfect optics; 2. Spherical aberration; 3. Astigmatism; 4. Turned edges; 5. Zonal defects; 6. Surface roughness (Adapted by the author from a diagram initially drawn by Philip S. Harrington).

If the images your telescope are producing are good and are satisfactory to you, you probably should not be reading this section anyway. On the other hand, if you are experiencing image problems you are virtually guaranteed to uncover the reason why using a well set up Ronchi test. Don't forget, the faults shown below become less visible when coarse bands (100 per inch, approximately) are used. There can be faults that are invisible with 100 lines but which still have deleterious effects on your image quality. This is the reason why it is always better to employ a grating with either 250 or 300 lines per inch. Little, if anything, escapes these fine bands.

What will you see? That's not easy to describe, so we've illustrated (Fig. 10.1 below), by means of some diagrams, the various faults that could be at issue. Indeed, combinations of these faults are usually evident in some 'scopes, so you need to evaluate what you are seeing in toto. The bands have been exaggerated in contrast to assist in the clarity of the explanation. You should see four parallel lines. Now, examine the defocused image inside focus (intra-focally). Are the lines straight both inside and outside focus? If so, then your optics are probably fine. Did you see curvature or a definite asymmetry in the pattern inside and outside focus? The basic idea is that you can use a computer to predict precisely what each Ronchi pattern should look like for different aberrations at various offsets in and outside of focus. After performing the test, you match what you've seen to the figure that best matches the predicted patterns as closely as possible. The diagrams below show the patterns that are characteristic of all the major aberrations that can plague your optics.

High Power Test

Unless you're a bit weird, or like me, test telescopes for a living, you'll spend most of your time looking at things in focus. It's fun to test the mettle of your Dob by cranking up the power on a night of good to excellent seeing. The obvious first choice is Luna, which can be observed equally well from a city or country setting. Look at the image of the first (or last) quarter Moon throws up when your 'scope is charged with a magnification of at least 50× per inch of aperture 30× per inch if your 'scope is 10 inches or larger. If the views are crisp and well delineated at these high powers then rest assured, you have a decent optic.

One enduring belief among amateur astronomers is that splitting close double stars make excellent optical tests. There's certainly more than a grain of truth to this, but it's not entirely accurate. The Dawes limit is an empirical result – and amply borne out in field tests – derived by the nineteenth-century English clergyman and amateur astronomer William

Rutter Dawes, who found that a telescope will not resolve two equally bright, sixth-magnitude stars if their separation in arc seconds is less than 4.56/d, where d is the aperture in inches. So, for example, a 3-inch 'scope should resolve pairs as close as 4.56/3 = 1.52 arc seconds and a 5-inch should do considerably better (0.91 arc seconds).

Now, according to Harold Suiter, an authority on the analysis of a star test, a 'scope with a ¼ wave of spherical aberration will split close doubles down to the Dawes limit, yet it'll give slightly softer, less well defined, lunar and planetary images compared to a 'scope corrected to say 1/6 of a wave or better. The atmosphere plays a fundamental role in limiting the resolution of fine detail. For example, a 12-inch Dob has a Dawes limit well below 0.38 arc seconds, yet you cannot split such a close pairing of double stars. Separations of 0.6 arc seconds are doable, but only on rare occasions. In many northern locations, the seeing almost never allows the largest Dobs to resolve to their theoretical limits. For this kind of work, it's best to stop down the aperture to better match the seeing to your aperture or use a smaller instrument.

That said, probably the best, most all-encompassing, and simplest test of your 'scope's optics is a high magnification examination of a bright planet. Jupiter is often the best. Do you want to know if your 'scope is optically sound? Point it at Jupiter when it's at least 30 degrees (the higher the better) above the horizon and away from any sources of heat on a calm, transparent night. Examine the planet at 30 to 50× per inch of aperture. Do you see a slightly flattened disk against an ink black background sky? Does the planet look off-white or maybe yellowish, crisscrossed by darker bands that vary in hue from milk chocolate brown to fawn? Is there structure within these bands? Can you see fragile ovals with odd colors merging with or separated from the bands? If you've answered "yes" to all these questions, then chances are you have a very nice optic. Of course, aperture will have a bearing on what you can expect to see. Planets are hard objects to image, and their low-contrast surfaces and/or atmospheric markings are most easily discerned when the 'scope is well corrected for Seidel errors as well as false color. If you're happy with the views your 'scope serves up, then that should be the end of the matter for you!

When all testing is done; how do you evaluate what you've seen? Are some aberrations worse than others? Is there in any sense a "shopping list" of optical defects that you can use to appraise the optical quality of your Dob? Well, to start with, any source of asymmetry is always going to be spotted first, since the eye is good at seeing it. So coma and astigmatism might be the first two. Some coma might be inherent in the optical design, so nothing will remove it; the same goes for astigmatism, but that can arise from strain in polishing or mounting the mirror and might be fixable. Some spherical aberration, if smooth, can be tolerated because at least one zone of the optics will be in focus. Roughness is a complex combination of many small errors and is the source of persistent poor contrast. For instance, you could have an objective with a wave-front with peak to valley (p-v) less than 1/10 wave but rough, with fast slope changes, which gives worse images than a smooth, ¼ wave P-V mirror.

The leading Dob manufacturers often sell their 'scopes complete with certificates of optical competency. Although this generally serves to reassure customers that the investment they have made has been justified, don't let spot diagrams, Strehl ratios, and intereferometry reports cloud your visual judgment. This author remembers testing a 8-inch aperture 'scope whose manufacturer claimed had a Strehl of 0.99 (virtually perfect) failed to split the famous double double in Lyra (Epsilon^1 and Epsilon^2) – a task more suited to a decent 4-inch Dob.

Most 'scopes sold today are designed by opticians, many of whom wouldn't know a planet from a star. They're not astronomers. They get on with what they do best, optimizing their designs for the maximum theoretical optical punch. Not surprisingly, these products provide textbook results when tested in the safe environment of the laboratory. After purchasing the 'scope, the excited amateur takes the 'scope outside, into an alien world that is often far removed from the cosy, climate-controlled environment in which it was first fabricated. Temperatures fluctuate wildly, creating tube currents, and winds induce vibrations. If that weren't enough, the act of moving the big primary mirror into different positions of the sky while moving from one object to the next warps the optics (if only a little). Net result: the 'scope fails to impress! Dobsonians, especially premium models, should be thoroughly field tested by manufacturers to ensure that they work in the field as they're supposed to. Telescope opticians should become stargazers, too! This bears repeating: *The eye is the ultimate arbiter when it comes to the discernment of optical quality.*

CHAPTER ELEVEN

Sketching and Imaging with Your Dob

One of the great virtues of owning a large Dobsonian telescope is that it can be used to plumb the depths of the deep sky, pulling out myriad nebulae, open clusters, and faint galaxies that skirt the great dark of interstellar space. The spectacular light-gathering power of even an 8-inch or 10-inch 'scope has inspired many amateurs to try their hand at astronomical sketching and astrophotography. This chapter explores some of the extraordinary achievements of amateurs using modest equipment and more than a little enthusiasm.

Astronomical sketching can be a very relaxing and rewarding way to capture what you see at the eyepiece and hone your skills as an observer. The human eye is a marvelous image reconstruction device, with a dynamic range unequalled by any imaging camera. Sure, CCD imagers can capture better shots of the Moon, planets, and deep sky objects by stacking lots of frames together, but there is something wonderfully compelling about being alone with a sketch pad and behind the eyepiece of a large aperture Dob. Sketching enjoys a degree of freedom that no CCD imager can experience – freedom from power chords, battery packs, temperature regulators and all the niggling things you need to tweak to get the right images. It is easy to see why more and more amateurs are turning their back on CCD imaging and instead are embracing the quiet simplicities of sketching. In the hands of masters like American amateurs

N. English, *Choosing and Using a Dobsonian Telescope*, Patrick Moore's Practical Astronomy Series 1, DOI 10.1007/978-1-4419-8786-0_11,

Jeremy Perez and Sol Robbins, using instruments of only 6- and 8-inch apertures, some wonderful artistic renditions of celestial show pieces can be created.

Pärt Veispak, from Tallinn, Estonia, is a physics student by day but a keen astronomical sketcher by night. His workhorse is a 12-inch Dob. He's been interested in astronomy since he was a kid but only took the hobby seriously in the last few years, investing in a small refractor. But Pärt soon decided to upgrade, so he opted for a 12-inch Dobsonian and took up the grand art of astronomical sketching. Here is how he described his routine: "My sketching workflow is relatively simple," he said. "I use an HB pencil, some A4 format paper of relatively decent quality, albeit it's not entirely too critical, a self-built height adjustable chair made out of spare planks, and a military lamp with red and green filters (fairly sure any red light will work, however, as long as it's dim enough). I also use a large notebook for my notes. When I've found the object I'm looking for, I use the notebook as support for the drawing paper. I hold the lamp in my left hand, along with the notebook and paper, which can be quite an exercise, and draw with the right. Holding the lamp is probably the most difficult part of sketching, so I occasionally use a variety of techniques to ease the process, such as hauling out a table as a footstool and balancing the notebook on my knee (Fig. 11.1).

In the field, all sketches I draw are relatively rough. I start by drawing whatever is the set of bright stars closest to the object in question, while using a low magnification, so it can act as a framework. I find that drawing the larger stars as empty circles helps speed up the process, and also helps in keeping them round. With star clusters, as I progress through the sketch, I start using higher magnifications to resolve tighter clumps, as well as reveal the dimmest stars. Getting them circular or point-like is a nigh impossible task, so I'm mostly focused on positions and relative brightnesses. Then, once the cluster itself is done, and depending on whether I think the cluster needs a context or not, I may draw patterns of brighter stars around the cluster, to make the borders of it, as well as its relative density easier to perceive. With nebulae and galaxies the technique is roughly similar, except after adding the bright and medium brightness stars, I add the nebulosity and only then proceed to the dimmest ones. I don't like attempting subtle shadings out in the field, so I just quickly sketch them, without worrying that the nebula more often looks like a series of rapidly drawn lines.

"With the planet Jupiter, the first thing I do is quickly divide the planet into zones with lightly drawn lines. The NEB, that faint belt that's where the SEB should be, and the polar regions. I then roughly fill the areas. It doesn't look smooth, of course, but there's no need for that at the eyepiece.

Fig. 11.1. Pärt Veispak's 12-inch GSO Dob (Photo used with permission).

Then I basically spend a lot of time finding the ideal magnification for what the weather allows, and staring at the eyepiece. For me, half an hour seems to be enough to get most of the detail I'm going to get. If I see belts or storms, I draw them in, as well as any denser knots or other aberrations on the surface. I clean the sketches up later, inside and under a proper light. I redraw stars that aren't point- or circle-like. Occasionally I just redraw the entire sketch on a new, and often better, sheet of paper. This is especially likely if I don't like the scale of the original sketch, so I can go proportionally upscale or downscale to achieve something that resembles the eyepiece view better. Nebulosities and the surface of Jupiter seem to require redrawing, since it's impossible to get the smooth gradients on the field. I use a cotton swab to blend the pencil strokes. This is a process that I repeat many times, adding more pencil strokes to areas that need to be darker and erasing in areas where the opposite is true (Figs. 11.2, 11.3, and 11.4).

Once the final drafts are done, technically one should scan them. However, I do not have one, so I have to use my digital camera. I try to set up fairly even lighting, and then use Photoshop™ and Pixinsight™ to get rid of the gradients, and invert the image for the DSO sketches. I've heard that using a cloudy sky can achieve even illumination, but I haven't tried it.

Does Pärt suffer from aperture fever, having used his 12-inch Dob so extensively over the last few years? "I have a Telescope Service branded GSO

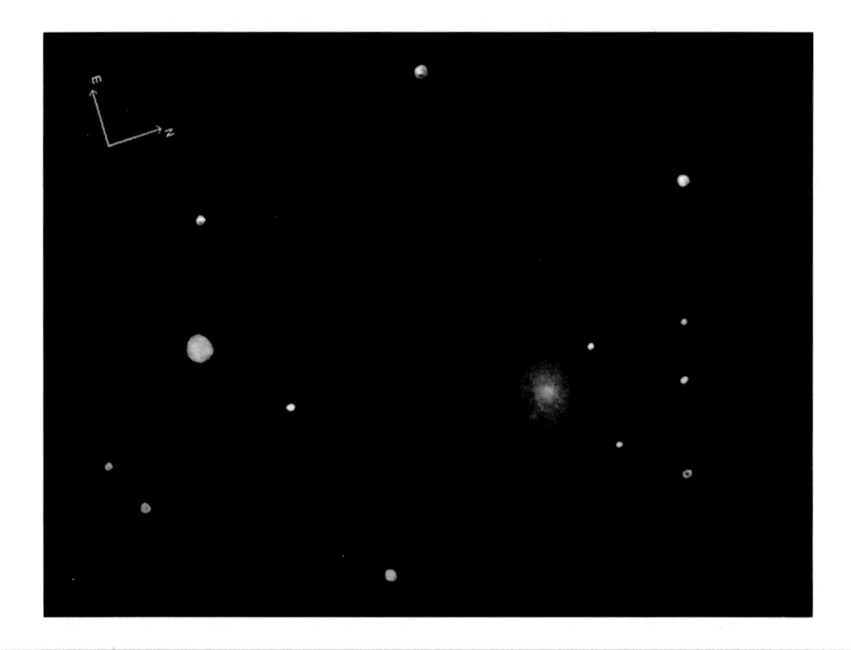

Fig. 11.2. NGC 404 as sketched by Pärt Veispak using a 12-inch f/5 Dob.

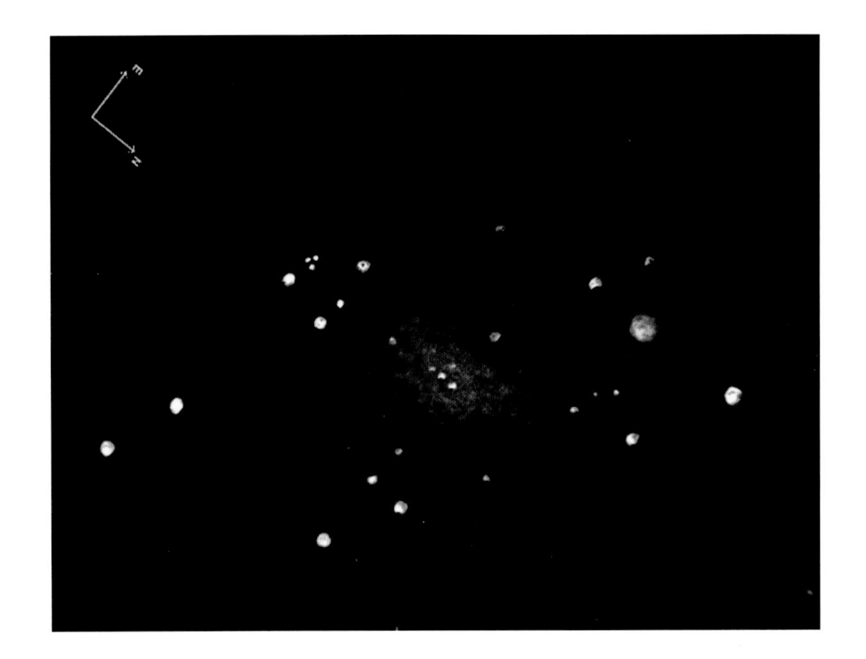

Fig. 11.3. NGC 6712 as sketched by Pärt Veispak using a 12-inch f/5 Dob.

Fig. 11.4. Jupiter, as drawn by Pärt Veispak using a 12-inch f/5 Dob.

300-mm solid tube. Of course, all Dobs are a bit of a work in progress, so it's had some faults I had to remedy, such as replacing springs, and some I have not yet gotten around to, such as getting new collimation screws. Aperture fever is an incurable disease in my opinion. I'll always wonder what I could do with a larger telescope, and would want one, should the situation allow for one. However, in many ways the symptoms have disappeared with this size. I almost never find myself thinking what I could see with something else when on the field. I'm often entirely delighted with the views I get. I also never feel intimidated by maps, and I know that, given the proper conditions and technique, I could probably see most objects marked on them. With my 50-mm refractor, I always felt that there was this veil draped across the sky, through which I could sneak a peek of some objects, whereas with the Dob, that feeling is gone. It's as if the entire sky has opened up. There's definitely no shortage of targets whatsoever. I believe I could get most of the NGC catalog at this aperture, with the exception of perhaps some of the very faintest galaxies, most of which are members of dim galaxy clusters. That being said, there's still a plethora of galaxy clusters within reach.

I should add that my usual observing sites are pretty good, despite living in the city. I spent the month of August in a small city of a pop of less than 5,000, where, when the conditions were good, I could see beyond magnitude 6.5. For the rest of the months, I usually stuff the Dob in a car and go to locations 50—70 km from where I live, so the limiting magnitude is around seven. That being said, a friend of mine has a 10-inch near the town, in what I believe is a yellow zone, and he has no problem seeing a plethora of objects.

Electronic Eyes on the Sky

Although traditionally, Dobs don't lend themselves to being camera friendly, innovations over the years have paved the way for these simply mounted instruments to produce stunning images of the Moon, planets, and fainter deep sky objects. Although neat snapshots of brighter objects can be taken through an undriven mount, some sort of tracking platform is highly desirable. As we have seen previously, a number of companies now offer high quality wooden and aluminum platforms for motorized tracking, while other models have powerful dual-axis motors built into the altitude and azimuth axes (or can be retrofitted if need be).

Taking snapshots with your Dobsonian couldn't be easier. You should be able to capture at least fairly decent shots of lunar vistas by simply aiming your camera phone into the high power eyepiece of your Dob. If you play around with this for a while, chances are good that you'll end up with some really impressive shots. For even better results, it's necessary to mate your Digital Single Lens Reflex (DSLR) camera directly to the focuser of the Dob via a suitable adapter (usually supplied with many econo-Dobs). To do this, you'll need to remove the camera lens and then screw in a so-called T mount that enables the camera to couple effectively with the 'scope's focuser. Since lunar and planetary shots require very short exposures, an undriven mount can be adequate enough for the task at hand. That said, you'll have to live with the inconvenience of having to nudge your 'scope along as the target moves across the field of view. That's why anyone serious about astro-imaging with their Dob requires a tracking mount.

Jonathan Usher from Wellington, New Zealand, has produced some amazing planetary images with very modest equipment. "My Dob is a Sky-Watcher 12-inch auto-tracking Dob," he said, "I haven't made any modifications to the 'scope itself at all (I do use a cooling fan though). Other equipment used is a manual Orion filter wheel with 1.25-inch RGB Orion filters, an Orion flip mirror, a 24-mm Panoptic eyepiece (for centering), a 2× "Shorty Plus" Orion Barlow, and a DMK21AU04 imager (mono; 640×480 frame size). I point the 'scope level and North, and turn on power (this is after collimation). Even though I can use the 'scope in "GoTo" mode, I don't typically do that – just start tracking and manually slew the 'scope towards the Solar System object I wish to image. After centering Jupiter within the 24-mm Panoptic, I use the Orion flip mirror to move to the camera and start capturing frames – typically at 30 frames per second in each of the R, G, and B channels. I typically end up with a bit over 1,000 frames per channel in the resulting AVI files. I process them using Registax and Ninox for alignment, and Astra Image for the final merged image.

I've attached one of my best Jupiter images (see below) from last year as an example result. I have found imaging Solar System objects with this setup to be a lot of fun, but not without significant challenges. Even very small breezes, which would not disrupt imaging using a Newtonian on a strong equatorial mount, make imaging Solar System objects at high resolution extremely challenging. This is because the surface area of the tube is quite high, but the mount isn't as solid/strong as, for example, my NEQ6 Pro mount that I have my 5-inch TeleVue 'scope mounted on. Added to that, there is considerable free play in the mount.

"With some work, I could likely tighten this up a bit. Lots of backlash/free play, and imperfect (albeit reasonable) tracking leads to lots of button presses to try and keep the object within the FOV – even for a 45-second imaging run. The scale I'm working at is pretty high (a bit under 0.2 arc-seconds per pixel), that Jupiter occupies a sizable part of the 640×480 frame size of the camera, and keeping any object in view is pretty challenging. I've moved to a motorized filter wheel for this year (Orion Nautilus) because using the manual wheel with this setup was very frustrating. The free play in the mount quite often resulted in the object being moved out of the FOV (by a lot) when the wheel was turned manually – even when I was very careful. I'm looking forward to using the motorized filter wheel for the first time soon. Field rotation can become an issue quite quickly, given the resolution I'm seeking to capture on images of Jupiter. I've done some manual rotation of R, G, and B channels to minimize the impact of this. This is fine for Solar System objects, but I can't imagine using this setup to try for any longer exposure imaging!

Even with all of these challenges, I still am very happy with the images I've been able to achieve of Jupiter. Clearly the optics of this 'scope are very good. He should be, as these are world-class caliber images (Fig. 11.5)!

Other astronomers have tried hooking up very low light video cameras to their tracking Dobs. With a 12-inch 'scope you can look

Fig. 11.5. Majestic Jupiter. Image shot through a 12-inch Sky Watcher Dob (Image credit: Jonathan Usher).

forward to seeing detail in the spiral arms of the Whirlpool Galaxy (M51) or the mind-boggling structural detail of M42 in the Sword of Orion. You can check out the fields of galaxies in the Virgo Cluster and see them in real time on a screen. Do you want to look inside the Eskimo planetary nebula? The list goes on and on. These cameras, when coupled with a tracking telescope, are hard to beat.

Of course, one really nice feature of these video cameras is that they enable you to record the video DVD (or tape if you prefer to view in analog mode), where you can later process them for posterity. In addition, they are arguably the ultimate tool in public outreach, since so many can people can see the amazing details a large Dob can unveil without the need for queues, ladders, and the like. Other amateurs are trying long exposure imaging using consumer digital video cameras and more specialized CCD cameras, such as the new Meade DSI camera.

As we heard from Jonathan Usher above, one of the key problems to overcome if you wish to image with a motorized alt-azimuth mount is that of field rotation. Field rotation is the apparent rotation of a celestial object in the field of view of a telescope during the course of the night. All objects in the eyepiece field or on the camera's image will move in arcs. It's usually either ignored or not noticed for visual observations, but you can't do that for long exposure shots. You see field rotation when you're using an alt-az mount or a misaligned equatorial mount. In these cases, all the stars will appear to move around the point or star that is being tracked. Field rotation will occur unless the mount is exactly aligned to counteract our planet's rotation.

Think about what happens when a constellation or the Moon rises, transits your local meridian (north–south line), and sets. That's exactly what field rotation does. Think about how the angle at which the various aspects of the lunar surface changes during the course of the night – first tilting to the east, then tilting to the west. Both examples show field rotation with your head and body acting like an alt-azimuth mount. Because of field rotation, the cardinal visual directions will change relative to the top and bottom of the eyepiece during the night.

Usually this is not important, but you should be aware of it. On photos, even if tracking on a star is perfect, unless the mount itself is perfectly polar aligned, there will be some degree of field rotation. Only the center will show pinpoint stars or sharp detail. Towards the edge of the field of view, all stars will be seen to trail round concentric arcs. If you're guiding a photograph and using a star off-center or outside the field of view, then the arcs will be concentric on the guide star. Because of field rotation, no alt-az mount is suitable for long-exposure astrophotography. Only a properly polar-aligned equatorial mount will eliminate field rotation. That said, software now exists to remove field rotation from your images. You can simply take 20 short exposures or more (to reduce any negative effects), rotate, stack, and process! In this way, you can simulate the effects of a single 60-min exposure. Imagine the amount of light that collects with a big Dob!

Dan Price from Ridgecrest, California, described his astro-imaging schedule with his giant backyard 'scope, a 25-inch Obsession Dob. "The easiest way to try prime focus astrophotography is just to attach the camera to the focuser and adjust it," he says. "I used SBIG's ST-10xme camera and a Feathertouch Focuser along with Optec, Inc's TCF-S focuser. If it won't focus, you may have to adjust the length of the truss poles. Most often this involves cutting them. The next improvement is to track. I used StellarCat's Servocat with Wildcard Innovations' Argo Navis for tracking. My next task was to correct for "jitter," since a Dob was never designed for high-precision tracking. I used SBIG's AO-7 to reduce this effect. Then, I had to correct for field rotation. I used Optec, Inc.'s Pyxis Rotator (2-inch). Then I discovered that vignetting and weight became a problem, with so many components attached to the 'scope. The solution for vignetting was to mount the camera in the UTA looking directly into the primary mirror, with the TCF-S focuser and the Pyxis Rotator BEHIND the camera. A "cage" for the camera had to be machined to support it (Figs. 11.6 and 11.7).

Fig. 11.6. The camera cage for the 25-inch Obsession Dob for prime focus astro-imaging (Image credit: Dan Price).

Fig. 11.7. Majestic M13 imaged through a 25-inch Dob, 6 × 30 s through RGB filters (Image credit: Dan Price).

All of this allowed images to be taken with exposures as long as 1 hour. But I did encounter some limitations:

1. Wind! Without an observatory and the sheer size of the 'scope, any breeze over 3 mph is a problem.
2. AO-7 tracking. The AO-7 tracking software was designed for an equatorial mount. This caused the tracking to fail once field rotation exceeded roughly 15–20 degrees.
3. Field rotation above 80 degrees elevation. The Pyxis was fine up to about 80 or 85 degrees elevation, when the field rotation became too high to follow. Not many deep sky objects go straight overhead, and if they do, this only means waiting less than an hour for them to progress to a lower elevation angle.
4. Weight. All that weight at the end of truss poles for a 25-inch f/5 parabolic mirror took a lot of counterweight. If the telescope was not on a perfectly level surface stiction (static friction) it became a problem.
5. Cables. I had as many as seven cables hanging off the front end of the telescope for power and communication with the various components.

Dan doesn't take himself too seriously when it comes to his imaging. "I never felt that my images approached the quality of modern amateur astrophotography," he says, "but it is fun to experiment and try something off the beaten path. There is certainly huge room for improvement in what I've done, but I haven't noticed anyone picking up where I left off."

If anything, Dan's modest work shows that large Dobs have enormous potential to produce a new breed of deep sky images delivered from the backyard (Figs. 11.7 and 11.8).

Astrophotography has well and truly embraced the Dobsonian telescope in ways that were frankly unthinkable just a few years ago. Advances in tracking and imaging software now allow the largest light buckets to take spectacular images of the heavens, and things can only get better. And that's where we're going next – to the future – to explore some of the avenues the Dobsonian revolution will take in the years ahead.

Fig. 11.8. The awe-inspiring Whirlpool Galaxy M51, as pictured in a single-shot 5-min exposure through a 25-inch Dob (Image credit: Dan Price).

CHAPTER TWELVE

Where Next, Columbus?

We have reached the end of our journey through the rich and varied world of the Dobsonian telescope. We have followed the movement from its inception over 30 years on the streets of San Francisco right up to the proliferation of designs commercial Dobs now embody. It is extraordinary that an eccentric man with a passion for putting together a large telescope from leftover junk would one day go on to inspire generations of amateur astronomers and stoke a multimillion dollar industry as a result.

Many experienced amateurs have a large Dobsonian telescope in their stable. And for good reason. In terms of light-gathering power and resolution, Dobs provide the biggest bang for your buck. But what does the future hold for the Dobsonian? Where can the revolution take us next? Well, for starters, publically owned 'scopes are probably going to get a whole lot bigger. Here's an inspirational story from California, where a group of amateurs have come together to raise funds for one of the largest amateur run telescopes in the United States, soon to open it eyes to the public.

N. English, *Choosing and Using a Dobsonian Telescope*, Patrick Moore's Practical Astronomy Series 1, DOI 10.1007/978-1-4419-8786-0_12,

The Story of LAT

If you thought that Dobs cannot conceivably get any bigger, you'd be surprised to know that the largest 'Dob' thus far built makes the 50-inch Orion monster look like a pussycat. We're talking of course, about the Large Amateur Telescope (LAT), a 71-inch (1.8-m) alt-azimuth mounted Cassegrain reflector, the creation of a non-profit educational organization called Group 70, comprised of people from diverse nationalities and age groups, united in a common purpose to make the beauty of astronomy available to as many ordinary people as possible. The project had its inception back in 1988, when Group 70 members got together for the first time with the goal of building the LAT. Upon completion, it will be the largest telescope in the world built by and for those who do not have access to large institution-run observatories. The project has ambitions to not just providing a large aperture telescope but to eventually offering related instrumentation and services to those in amateur, professional, and educational fields of astronomy.

The glass mirror blank, 71 inches across, which will form the heart of the optics of the LAT, was actually obtained from the University of Tasmania, Australia. Constructed from tried and trusted Corning Pyrex glass, the mirror was originally cast in 1938 as a backup blank for the 48-inch Schmidt camera in the observatory atop Mount Palomar. The blank is currently in the safe hands of Elliott Laboratories in Fremont, California. Group 70 will operate the 'scope as an f/10 Nasmyth Cassegrain, and which will be capable of doing state-of-the-art CCD imaging and research, particularly in spectroscopic and photometric studies.

Like the Faulkes Telescopes already established in Europe, it is planned for the LAT to be fully operational remotely via the by worldwide web, thereby enabling individual amateurs, astronomy clubs, schools, and other educational institutions access to this super large telescope. The LAT will be located in California's "Dark Sky Corridor," a 200-mile long stretch of ridges in the coastal mountain ranges along the Pacific Ocean, lauded for the pristine skies it presents. Several acceptable sites are now being evaluated. The bylaws stipulate situating it within 4 h driving time of Cupertino, California, but committee members are still not certain where they will place it. That said, the site selection has now been whittled down to just three locations. The first is Chew Ridge, a great dark sky location but not easily accessible – as is typical of all great locations. The other two sites beng considered are either Mount Hamilton or Mount

Wilson, both of which are much more accessible but pay the price of having more light pollution.

Group 70, as a non-profit educational organization, is governed by a board of directors who meet monthly to act on these goals. The LAT will be operated by volunteers who give their time, talents, and resources to getting the project off the ground and running. These include manual work on the telescope and it components, soliciting grants and other funding, and planning for the future of the observatory, telescope, and organization.

Its dimensions are pretty hard to imagine unless you're standing beside it. The tube will measure 7 feet across and 25 feet high. The enormous mirror will weigh in at 2,420 pounds. Incidentally, it was initially found to be over corrected, so steps were taken to bring it back to a spherical shape again. At the time of writing, the final figuring to exact its parabolic shape (at f/2.8!) is being done by volunteers who can run the grinding machines. Most work gets done on Saturdays, when LAT members have some free time from their busy life. The mirror will be mounted on a 27-point flotation cell.

Though a Cassegrain design, Group 70 have opted for an alt-azimuth mounting for it for a number of reasons. For one thing, it's lighter and less expensive. It's also easier to operate that way and takes up less room. Like many Cassegrains, the figuring on the secondary mirror will commence once the figure of the primary is accurately known and assessed. Once completed the LAT will be the fourth-largest telescope in California, eclipsed only by the 200-inch Hale reflector on Palomar, another 120-inch atop Mount Hamilton, and the 100-inch Hooker on Mount Wilson. If everything goes to plan, the LAT – arguably the world's largest 'Dob' – will be fully operational by 2014.

Going Faster?

But what of the average Joe, sitting in his or her back garden? How will the Dobsonian revolution unravel itself for him in the coming years and decades? We've already seen that Rick Singmaster, founder of Starmaster Telescopes has taken the large Dob in one direction, by lowering the focal ratio of the mirror to f/3.3 on his largest instruments thereby shortening tube length and allowing even large telescopes such as his 22-inch Super FX, without the need for a step ladder. Mike Lockwood has made a 20-inch f/3 Dobsonian. So why not go even lower? After all, the observatory class instruments have huge, ultrafast mirrors. For example, the 10-meter Keck telescope atop Mauna Kea, Hawaii, is f/1.75! That trend, most likely, is set to continue.

Getting superfast mirrors is possible. Yes, it takes money and engineering (both in large measure) but some of the more talented ATMers will be able to make it work.Whether or not it will become a large movement depends upon how many people will pay the money and want to become engineers to keep everything in tolerance. And those tolerances will have to be tight. Making an optically sound f/2.5 mirror is no mean feat. That said, as the early part of the Dobsonian revolution showed us, there are many who are not bothered by stars that look like round blobs instead of tight pinpoints of light.

Many are inclined to believe that the Dobsonian revolution is already in its heyday. For one thing, enormous mirrors take too long to reach thermal equilibrium, and by "too long" we mean considerably longer than the time an average observer can reasonably set aside for an evening. Enormous mirrors of the future will still be quite heavy, despite having ultra low density honey comb structures. While one can 'justify' the amount of weight one can reasonably lift and carry over short distances, it quickly becomes a tedious and exhausting affair.

What's more, judging by online threads here and elsewhere, fine collimation still eludes many amateurs when using common focal ratios of around f/4 or f/5. That problem becomes much more challenging at f/2.5! To have the slightest of chances of actually holding collimation as the 'scope slews across the sky, every component – especially the optics – that makes up the mechanics of the 'scope will need to be of premium quality, and that means the price goes up.

At f/2.5, most reasonable eyepieces will begin to show major distortions. For a start, they would generate very large exit pupils. For example, even a 12.5-mm focal length eyepiece would yield a 5-mm exit pupil! But that's an unnecessary evil, especially for more mature observers, as it brings you right into the range where astigmatism becomes a problem. If you elect to use a 3-mm exit pupil as the largest that is consistent with lack of natural eye astigmatism and wish to use eyepieces in the range 15 mm to 20 mm, your f/ratio is going to work out to 4 or 5 at the fastest. Even at f/4, a 24-mm eyepiece gives a 6 mm exit pupil!

In addition, a Dob operating at f/2.5 will certainly require a coma and astigmatism corrector with a power exceeding that delivered by any existing corrector on the market. It will require perfect optics, and ultra-solid mechanics that are, at the same time, portable. For these reasons, I don't think any mass market/low cost producers are going to offer a Dob with a superfast f/2.5 mirror any time soon.

In the next decade or so, the real advances will be embodied in the marriage of electronics and eyepieces. It's a revolution that has yet to reach the masses. Think how exciting it would be if something remotely resembling a real-time color eyepiece was developed – the technological successor to the Collins I cubed intensified eyepiece with a good photoelectric response across the visual spectrum. It will be pricey, too (at least initially), but it might just enable smaller, easily portable, low and intermediate F ratio reflectors to have the light grasp of a much larger 'scope. Now wouldn't that be nice?

Another way around the problem of building ultrafast mirrors is to design eyepieces that can correct for the aberrations they induce. In a sense, the feasibility of such an approach has already been demonstrated with mirrors down to f/4. The optical innovations brought to market by Al Nagler with his Paracorr and revolutionary eyepieces shows that it can be done on still faster optical systems. If you throw enough cash and human resources into the project, then amateurs astronomers could be free to enjoy 30-inch diameter backyard behemoths without the use of awkward ladders.

The Conical Mirror

There are many ways to innovate, but few truly raise the bar in terms of performance. What if you could build large aperture Dobs with mirrors that give you first-rate optics but are 50 percent lighter, cool off much faster, are easier to mount inside the tube, and hold collimation better. Enter the world of the conical mirror. As a design concept, the conical mirror – so-called because of its cone shape – isn't exactly new. Indeed, it has long been used in the primary mirrors of commercial Schmidt Cassegrains and Maksuov Cassegrains. Most telescope makers have never considered it as a viable alternative to the conventional 'cylindrical' mirror because the blanks are more difficult to produce. But once you get over that hurdle – and it's possible – you have an entirely new kind of mirror for the Newtonian. Because of its conical shape, these mirrors can be made up to 60 percent lighter than a conventional mirror of the same aperture. In addition, with a conical cross section, its apex measures only 25 percent of its diameter, so it's considerably easier to mount rigidly at the base of the optical tube.

The conical mirror also has some noteworthy thermal properties. Conventional mirrors can take a long time to acclimate, as we've seen. Because of its shape, a conical mirror will have a much larger percentage of its surface area exposed to the surrounding air, as well as to any installed cooling falls. Indeed, because the mirror is thinner at the edge and cools down at a faster rate than a full thickness mirror, it doesn't produce that classic case of 'overcorrection' as the mirror cools down to ambient, so you can get nice images faster. Another notable advantage of the conical mirror is that it does not require a complex floatation cell. Nor is there any need for mirror clips that could potentially distort images if tightened too much. All said, the conical mirror retains collimation much better than standard models.

Master optician Bob Royce based at Northford, Connecticut, has developed the conical mirror for his line of reflecting telescopes. He explained why he has invested so much of his time and energy into developing these products. "I have pretty much invested my reputation in conical mirrors for all types of telescopes, especially Newtonians," he says. "In 2003, I began to sell them for Newtonians as well as for the more traditional use in Cassegrain types. My contribution was the development of a mounting system that would make conical mirrors suitable and practical for Newtonians. Initially, I sold only 12.5-inch mirrors completely perforated through the center and held with a traditional O-ring hub. Later, in 2004, I began to make and sell 8-inch and 10-inch mirrors of my own mold design, having a new mounting method that preserved the face of the mirror, the so-called continuous face system. In 2006 this was expanded to a 12.5-inch size made from my own mold design that saves a bit on weight. The 14.5-inch and 16-inch Newtonian mirrors are still held with a through-core hub. Conical mirrors for my 8-inch through 12.5-inch Dall-Kirkhams no longer use hubs with O-rings but are actually potted into hubs using a special RTV (Fig. 12.1).

Bob was asked what the advantages of conical mirrors were. "These mirrors have numerous advantages," he continued, "the three most significant being: rapid thermal stabilization, light weight, and ease of mounting. Secondary advantages include the elimination of mirror clips (which are actually quite damaging) and figure-controlled thermal stabilization. This last feature is a serendipitous advantage that apparently results from the conical shape itself. The mirror does not necessarily go into significant over correction during cooling. This is very apparent during fabrication, where stabilization occurs more rapidly than with a standard mirror and the figure does not experience nearly as much change during and after transition.

"As far as mounting is concerned for Newtonian mirrors you need only drill a hole through a tip-tilt plate and mount the mirror by screwing it down with the hardware provided. That's right, just screw it down! And you don't need to make a floatation system cell as is the case with

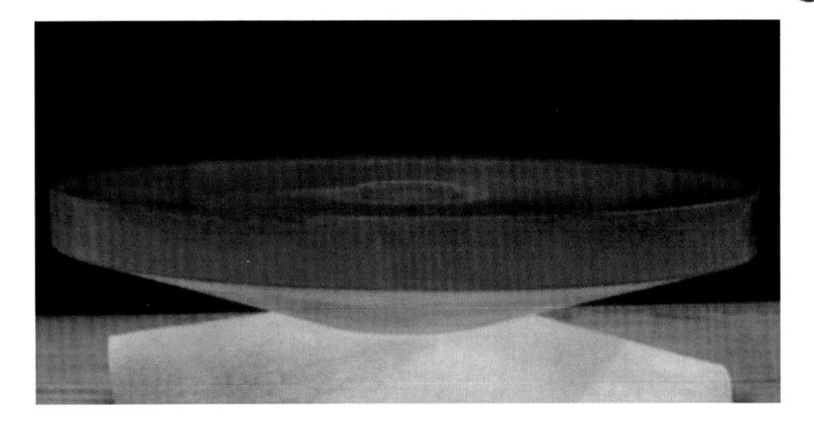

Fig. 12.1. The conical mirror (Image credit: Bob Royce).

thin mirrors. Some people have inquired (and incorrectly assumed) that 'the nut' in the center of the mirror will cause a certain amount thermal or mechanical distortion and that these mirrors are somehow 'compromised'. This is not the case. The hub is held in place using a barrier layer of RTV that absorbs any thermal discontinuity.

If one looks at a cross-sectional representation of a mounted conical mirror and a standard mirror, what becomes apparent is that the conical mirror allows for a free circulation of air around the mirror. The mirror is not confined in any way in a cell that holds warm air and inhibits the cooling process. The entire back as well as the front of the mirror is free to radiate heat.

As to potential use for imaging, these mirrors have one immense advantage – they simply do not move. A standard mirror must of necessity not be held tightly or they will bend in some way due to thermal changes. Clever devices have been designed to reduce this defect, but all standard mirrors shift, even if just a little.

You can see why Bob Royce is pushing the boat out with his conical mirror. It promises to raise the observing bar for the backyard hobbyist by increasing the time spent observing and not waiting for cool down, extending your observing mileage under the starry heavens. Who knows, one day it might well a standard feature in all commercial Dobs. But that's the beauty of it all. No one can portend the future.

As these words are written, John Dobson is just ending his 96th trip around the Sun. And though his public engagements are not as frequent as they used to be, his thoughts and ideas have been spread far and wide by the Internet. I think it is fair to say that Dobson's astronomical

Fig. 12.2. Fire starter: John Dobson.

evangelism galvanized a movement that, in a real sense, was 'waiting to happen.' Perhaps he's easy to dismiss as a maverick character with peculiar ideas. But somehow, directly or indirectly, Dobson started a movement that gave ordinary people like yours truly the opportunity to look through very powerful telescopes and discover for themselves the joys of large aperture observing. On behalf of all those who took time out of their busy lives to contribute to this book, we'd like to wish John Dobson many more years of glorious existence.

Appendix A

References

The History of the Reflecting Telescope

King, H. C., *The History of the Telescope* (1955), Dover.
Hoskins, M. (ed.), *The Cambridge Concise History of Astronomy* (1999), Cambridge University Press.

Newtonian Optics

Kriege, D., & R. Berry, *The Dobsonian Telescope: A Practical Manual for Building Large Aperture Telescopes* (1997), Willmann-Bell Inc.
Bell, L., *The Telescope* (1981), Dover.
Rutten, H. G & M. van Venrooij, *Telescope Optics, a Comprehensive Manual for Amateur Astronomers* (2002), Willmann-Bell Inc.
Sidgwick, J. B., *Amateur Astronomer's Handbook*, (1971), Dover Publications.
Suiter, H. R., *Star Testing Astronomical Telescopes* (2009), Willmann-Bell.
Kitchin, C. R., *Telescopes and Techniques* (1995), Springer.
Pfannenschmidt, E., *Sky & Telescope*, April 2004, pp 124–128.

N. English, *Choosing and Using a Dobsonian Telescope*, Patrick Moore's Practical Astronomy Series 1, DOI 10.1007/978-1-4419-8786-0,

Observational Astronomy Books

Ridpath, I & W. Tirion, *Star & Planets* (2007), Collins.
Ridpath, I. (ed.) *Norton's Star Atlas and Reference Handbook* (2003), Longmann.
R. Consolmagno, G., & Davis, D. M., *Turn left at Orion* (2000), Cambridge University Press.
P Argyle, B.(ed.), *Observing and Measuring Double Stars* (2004), Springer.
Haas, S., *Double Stars for Small Telescopes* (2006), Sky Publishing.
Handy, R., *et al, Astronomical Sketching: a Step by Step Introduction* (2007), Springer.
Mullaney, J., & W. Tirion, *The Cambridge Double Star Atlas* (2009), Cambridge University Press.
Price, F.W., *The Planets Observer's Handbook* (2000), Cambridge University Press.

General Guides to Equipment

Backich, M., *The Cambridge Encyclopedia of Amateur Astronomy* (2003), Cambridge University Press.
Dickinson, T., & A. Dyer, *The Backyard Astronomer's Guide* (2008), Firefly Books.
Harrington, P. S., *Star Ware* (2007), Wiley.
Mobberley, M., *Astronomical Equipment for Amateurs* (1998), Springer.

Astrophotography

Ratledge, D., *Digital Astrophotography: the State of the Art* (2005), Springer.
Reeves, R., *Introduction to Digital Astrophotography*(2005), Willmann-Bell.

Recommended Monthly Periodicals

Astronomy Now. The leading and best established British astronomy magazine.
BBC Sky at Night magazine. A popular choice among UK amateurs.
Sky & Telescope and *Astronomy* magazines. For up to the minute reviews and all the latest news, with a North American slant.
Astronomy & Space. Published by Astronomy Ireland; a good resource for all things astronomical in the Emerald Isle.

Websites of Interest to Dobsonian Enthusiasts

Vladimir Sacek's excellent website on telescope optics: http://www.telescopeoptics.net.
A link to the sensational San Francisco Sidewalk astronomers and some interesting information about John Dobson and his mission: http://www.sfsidewalkastronomers.org/index.php?page=contact.
Roger Ceragioli's Refractor Construction Page: http://bobmay.astronomy.net.
www.excelsis.com. A surprisingly good review site for telescopes and astronomical accessories.
www.cloudynights.com. A superb resource to the astronomical community.
www.Astromart.com. A marketplace for astronomical ware and equipment reviews.
www.iceinspace.com.au. An excellent online resource for Australian amateurs.
www.Scopereviews.com Ed Ting's eclectic website featuring mini-reviews on some nice Dobs of older pedigree.

Appendix B

Useful Formulae

Eyepiece magnification = focal length of the objective/focal length of eyepiece
Field of view (angular degrees) = Apparent field of view of eyepiece/ eyepiece magnification (approximate). A more accurate formula is given by:

$$(\text{eyepiece field stop diameter} / \text{focal length of telescope}) \times 57.3$$

Focal ratio of telescope = Focal length of telescope/objective diameter
Exit pupil: telescope aperture (mm)/magnification of eyepiece.
Depth of focus:
$\Delta F = +/-4\lambda F^2$, where λ is the wavelength of light and F is the focal ratio of the telescope.
Angular measurement: 1 angular degree = 60 minutes of arc (60′) = 3,600 seconds of arc (3,600)
Limiting magnitude of a telescope = 6.5 – 5logd + 5lod D, where d is the diameter of the observer's pupil when dark adapted and D is the aperture of your telescope.

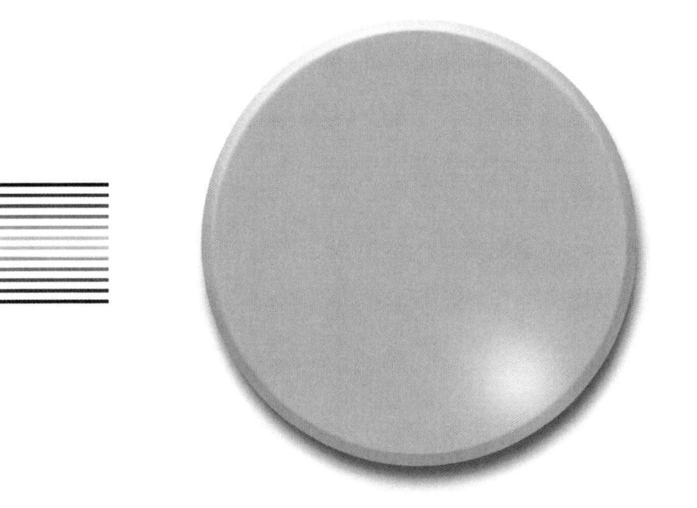

Appendix C

Glossary

Airy disk The disk into which the image of a star is spread by diffraction in a telescope. The size of the Airy disk limits the resolution of a telescope.

Alt-azimuth A type of mount, like a simple photographic mount that allows you to make simple movements from left to right (azimuth) and up and down (altitude).

Antireflection coating The application of a very thin layer of a substance (i.e., magnesium fluoride) to the surface of the lens that has the effect of increasing light transmission and reducing internal reflections in the glass.

Astigmatism An aberration that occurs when there is a difference in the magnification of the optical system in the tangential plane and that in the sagittal plane.

Autoguider An electronic device that makes use of a CCD camera to detect guiding errors and makes automatic corrections to the telescope's drive system.

Autotracking The ability of a Dob mount to track objects as they move across the night sky.

Barlow lens A concave achromatic lens with negative focal length, used to increase the magnification of a telescope.

Collimation The process of ensuring that all the optical elements in a Newtonian are perfectly in line with each other for maximum performance

Coma An aberration that causes a point object to be turned into a pear or comet-shaped geometry at the focal plane, and which most commonly manifests itself off-axis.

Conical mirror A cone-shaped mirror that can be mounted more easily and cools down faster than conventional 'cylindrical' mirrors.

Depth of focus A measure of how easy it is to attain and maintain a sharp focus. The larger the focal ratio of your 'scope, the greater its focus depth.

Declination (Dec) Declination in astronomy is comparable to geographic latitude, but projected onto the celestial sphere. It is measured in angular degrees north or south of the celestial equator.

Diffraction A wave phenomenon that occurs when waves bend or distort as they pass around an obstacle.

Dispersion The tendency of refractive materials (i.e., a lens or prism) to bend light to differing degrees, causing the colors of white light to separate into a rainbow of colors.

ED Short for Extra low Dispersion, usually referring to glass that focuses red green and blue light more tightly than a regular crown flint objective and resulting in better color correction.

Extrafocal Outside focus.

Eye relief The distance from the vertex of the eye lens to the location of the exit pupil.

Focal length The linear distance between a lens and the point at which it brings parallel light rays to a focus.

Focal ratio The focal length of a telescope divided by its aperture.

Fresnel rings The set of diffraction seen around stars just outside and inside focus.

Intrafocal Inside focus.

Lazy Susan A type of cradle-style alt-az mount common to Dobsonians.

Magnification The factor by which a telescope makes an object larger.

Multicoated Where lenses are antireflection-coated, with more than one layer of coatings.

Parabolic mirror The main mirror in a reflecting telescope that usually has a parabolic shape.

Right Ascension (RA) Right ascension is the celestial equivalent of terrestrial longitude. Both right ascension and longitude measure an angle that increases toward the east as measured from a zero point on an equator, which, by convention, is the first point of Aries.

Rochi test A method of determining the surface shape/figure of a mirror used in reflecting and other types of telescope.

Secondary mirror The flat mirror that directs light from the primary mirror of the telescope into the eyepiece.

Spherical aberration The inability to focus rays of light emanating from the center and edges of a lens at a single point in the image plane.

Spherical mirror A mirror with a spherical figure usually only found in smaller reflecting telescopes.

Strehl ratio A measure of optical quality that measures how much an optic deviates from perfection. A Strehl ratio of 1.0 is the best one can attain.

Truss tube A type of open-tube Dobsonian design that uses trusses (poles) to keep the optical components in precise alignment.

Turned edge An aberration that occurs when the edge does not end abruptly but curls over gradually, starting from about 80 percent of the way out from the center of the mirror.

Zonal errors Localized defects that arise during the figuring and polishing of optical mirrors.

Index